Advances in Industrial Control

Springer
London
Berlin
Heidelberg
New York
Barcelona
Hong Kong
Milan
Paris
Santa Clara
Singapore
Tokyo

Other titles published in this Series:

Advanced Control of Solar Plants
Eduardo F. Camacho, Manuel Berenguel and Francisco R. Rubio

Control of Modern Integrated Power Systems
E. Mariani and S.S. Murthy

Advanced Load Dispatch for Power Systems: Principles, Practices and Economies
E. Mariani and S.S. Murthy

Supervision and Control for Industrial Processes
Björn Sohlberg

Modelling and Simulation of Human Behaviour in System Control
Pietro Carlo Cacciabue

Modelling and Identification in Robotics
Krzysztof Kozlowski

Spacecraft Navigation and Guidance
Maxwell Noton

Robust Estimation and Failure Detection
Rami Mangoubi

Adaptive Internal Model Control
Aniruddha Datta

Price-Based Commitment Decisions in the Electricity Market
Eric Allen and Marija Ilić

Compressor Surge and Rotating Stall
Jan Tommy Gravdahl and Olav Egeland

Radiotherapy Treatment Planning
Oliver Haas

Feedback Control Theory for Dynamic Traffic Assignment
Pushkin Kachroo and Kaan Özbay

Control and Instrumentation for Wastewater Treatment Plants
Reza Katebi, Michael A. Johnson & Jacqueline Wilkie

Autotuning of PID Controllers
Cheng-Ching Yu

Robust Aeroservoelastic Stability Analysis
Rick Lind & Marty Brenner

Performance Assessment of Control Loops: Theory and Applications
Biao Huang & Sirish L. Shah

Data Mining and Knowledge Discovery for Process Monitoring and Control
Xue Z. Wang

Tan Kok Kiong, Wang Qing-Guo,
Hang Chang Chieh
with Tore J. Hägglund

Advances in PID Control

With 91 Figures

Tan Kok Kiong, PhD
Wang Qing-Guo, PhD
Hang Chang Chieh, PhD
Department of Electrical Engineering, National University of Singapore,
10 Kent Ridge Crescent, Singapore 119260
Tore J. Hägglund, PhD
Department of Automatic Control, Lund University, Box 118,
SE-221 00 Lund, Sweden

ISBN 1-85233-138-0 Springer-Verlag London Berlin Heidelberg

British Library Cataloguing in Publication Data
Advances in PID Control. - (Advances in industrial control)
1.PID Controllers
I.Kiong, Tan Kok
629.8
ISBN 1852331380

Library of Congress Cataloging-in-Publication Data
Advances in PID control / Tan Kok Kiong ... [et al.].
p. cm. -- (Advances in industrial control)
Includes bibliographical references and index.
ISBN 01852331380(alk. paper)
1. PID controllers. I. Tan, Kok Kiong, 1967- . II. Series.
TJ223.P55A38 1999 99-30200
629.8--dc21

Typesetting: Camera ready by authors
Printed and bound by Athenæum Press Ltd., Gateshead, Tyne & Wear
69/3830-543210 Printed on acid-free paper SPIN 10703375

Advances in Industrial Control

Professor Dr -Ing M. Thoma
Institut für Regelungstechnik
Universität Hannover
Appelstr. 11
30167 Hannover
Germany

Professor H. Kimura
Department of Mathematical Engineering and Information Physics
Faculty of Engineering
The University of Tokyo
7-3-1 Hongo
Bunkyo Ku
Tokyo 113
Japan

Professor A.J. Laub
College of Engineering - Dean's Office
University of California
One Shields Avenue
Davis
California 95616-5294
United States of America

Professor J.B. Moore
Department of Systems Engineering
The Australian National University
Research School of Physical Sciences
GPO Box 4
Canberra
ACT 2601
Australia

Dr M.K. Masten
Texas Instruments
2309 Northcrest
Plano
TX 75075
United States of America

Professor Ton Backx
AspenTech Europe B.V.
De Waal 32
NL-5684 PH Best
The Netherlands

SERIES EDITORS' FOREWORD

The series *Advances in Industrial Control* aims to report and encourage technology transfer in control engineering. The rapid development of control technology has an impact on all areas of the control discipline. New theory, new controllers, actuators, sensors, new industrial processes, computer methods, new applications, new philosophies..., new challenges. Much of this development work resides in industrial reports, feasibility study papers and the reports of advanced collaborative projects. The series offers an opportunity for researchers to present an extended exposition of such new work in all aspects of industrial control for wider and rapid dissemination.

PID or three term controllers are widely used in industry. In a typical papermill it had been reported that there are about 2000 loops and 98% of these are PID control loops. This leaves forty loops for the application of advanced control. Thus cost effective and efficient technology to tune the PID loops is very important.

Books and monographs devoted tc the development of PID control are relatively few. Many control books, even process control books, will only have a chapter or two on PID control. The Astrom and Hägglund book *PID Controllers: Theory, Design and Tuning*, (ISBN 1-55617-516-7) is one notable contribution to the PID bibliography. However, research activities in the methods and technology of PID tuning are demonstrably well and truly alive. Much of the research has been centred around specific groups in Singapore, Taiwan, Australia and Canada. Recently the Advances in Industrial Control series published the monograph *Autotuning of PID Controllers* (ISBN-3-540-76250-7) which reported the work of Cheng-Ching Yu and his colleagues in Taiwan.

This new monograph by K.K. Tan, Q.-G. Wang, C. C. Hang and T. Hägglund provides a good presentation of the state-of-the-art in PID control, and highlights many of the innovations made to the subject by the Singapore research group. It can be viewed as a very useful status report on some PID control research as we go through to the next Millennium. The key tuning areas tackled are classical designs, automated tuning and multi-loop control. Practical issues are given a chapter, and a very interesting Appendix presents details of a sample of industrial process PID controllers. Industrial and process engineers will find this a very accessible introduction to the fundamentals and some advanced topics in PID. The academic community will undoubtedly be interested to discover the recent progress made in PID control design research. Thus, this monograph by Dr. Tan and his colleagues is a timely and welcome entry to the Advances in Industrial Control Monograph series.

M.J. Grimble and M.A. Johnson
Industrial Control Centre
Glasgow, Scotland, UK

PREFACE

PROPORTIONAL-INTEGRAL-DERIVATIVE (PID) is a familiar term of high significance to many engineers, technicians and other practitioners involved in automatic control systems. Controllers of the PID type have existed for more than fifty years. Today, PID controllers can be found in virtually all control systems, with applications ranging from process conditions regulation to precision motion control for assembly and process automation. This is not surprising since the reliability of the PID controllers has been field proven by decades of successful applications. The wide acceptance and massive support from control engineers all over the world ensure they have remained the single most important tool in the control toolbox. On the other hand, the benefits derived from the implementation of more complex and advanced control systems have never been quite obvious, despite the remarkable theoretical advances and breakthroughs which have been achieved so far. This is especially true under practical operating conditions of industrial control systems, where harsh and varying environmental conditions, imperfect system models, generally modest operator expertise available, and short breakdown recovery tolerance usually render the application and implementation of advanced control very difficult.

Far from becoming stagnant and obsolete, the research and development efforts for the evergreen PID controllers have been undergoing a resurgence in recent years. A lot of effort has been devoted to capitalizing on the advances in mathematical control theory while still essentially retaining the decades-old classical control structure. The new generation PID controllers developed are able to demonstrate very good control characteristics such as higher performance robustness and tighter control performance. The application base of PID controllers has also been further expanded, with these controllers now being applied effectively to systems and processes never before possible under traditional PID control. A lot of research effort has also been put into giving a higher level of operational autonomy to PID controllers. Many of these research works have already been translated into new and useful functions of industrial control products, such as those which enable automatic tuning and continuous re-tuning of PID control parameters. These features have been instrumental in reducing the reliance on long and tedious manual tuning procedures, thereby achieving cost savings in terms of manpower and product quality, and contributing to overall higher productivity in modern manufacturing and automation systems.

The book covers in a unified, comprehensive and complete treatment, the important knowledge relating to the background, application, design, implementation and advances of the PID controller. It is intended to be useful for a wide spectrum of readers who are interested in control engineering work using PID control, ranging from practising technicians and engineers to undergraduates and senior graduates. For the practising reader, it will serve as a ready handbook for reference and easy retrieval of useful operational information and direct formulas relating to PID control. For the more academic and research-oriented readers, the book contains thorough theoretical treatment of major results and new advances in PID control. It is indeed a key objective of the authors to organize and write the book as systematically as possible with the ambition that the book will become an important written document devoted to this very important controller of all time.

In what follows, the contents of the book will be briefly reviewed. Chapter 1 provides an introduction to the PID controller. The evolution of the controller is traced from the time of its inception through the years of mechanical, pneumatic and hydraulic instrumentation and devices, to the present digital micro-processor dominant era. Topics fundamental to the PID controller will be presented. These include the structure and construction of the PID controller, the constituent PID components, the various PID permutations and application base, and the different forms of PID controller available.

Chapter 2 is mainly a systematic review of the classical PID controller tuning methods. It first presents the key objectives of a PID tuning exercise and the main principles behind the classical tuning methods. The common rules of thumb are covered in detail in the chapter, elaborating on the effects of the variation of each of the PID components on the control performance in terms of speed and stability. A deliberate and concerted focus is devoted to illustrate the well-known *Ziegler-Nichols* methods in both the time and frequency domain, and the relationship between these variants. A generalized modified form of *Ziegler-Nichols* tuning is also presented, showing the control enhancements possible with modest modification to the tuning rules. Other classical tuning methods covered include the *Cohen-Coon* method, the *Stability Margin* method and the *Tyreus-Luyben* method.

Chapter 3 examines the limitations of conventional PID control when applied to more complex problems. It is well-known that when good performance is necessary, conventional PID control is ineffective when applied to certain notorious classes of systems, such as high-order and time-delay systems, poorly damped and unstable systems, non-linear, time-varying systems and uncertain systems. Recent work devoted to resolving these limitations, and thereby expanding the scope of PID control applications, is assembled and organized in a systematic manner in the chapter. Advanced control design techniques, including those based on *Pole Placement*, *Dominant Pole Placement*, *Exact Gain and Phase Margin*, *Linear Quadratic Control*, and *Composite PI/Adaptive Control* are introduced in detail. The chapter elaborates on new and refreshing approaches to the use of these methods for the tuning of PID control parameters to achieve good control characteristics while still essentially working on a simple PID control structure.

In large modern process plants, especially where the number of loops to be tuned is tremendous, it is no longer sufficient that process controllers be well tuned, but they must lend themselves to be tuned, re-tuned or self-tuned with minimal intervention from process operators. Automatic tuning and adaptation of PID controllers have been successfully applied to industrial process control systems in recent years. Chapter 4 provides a survey of the different automatic and self-tuning methods developed for the PID controller, including transient analysis, frequency response analysis and pattern recognition methods. Many of these methods are based on the initial extraction of a process model from a suitably-designed experiment, and subsequent tuning of the controller with respect to that model. A robust step response method which addresses many of the noise-related deficiencies of step-type tests is described. Particular attention is devoted to relay feedback automatic tuning and its various refinements since its formulation in the early 1980s by Professor Karl Astrom. The chapter also presents modified relay setup and a new on-line version of relay auto-tuning applied to controller-stabilized processes. These largely expand the application domain of the basic relay method to more difficult control scenarios with different requirements. The chapter also covers the application of digital signal processing techniques to naturally occurring transient signals in the closed-loop for on-line continuous self-tuning of PID control parameters.

Chapter 5 addresses a new form of PID control for multivariable processes. Unlike conventional approaches which either ignore inter-loop coupling or work towards eliminating it with decouplers, the method views each loop as an independent equivalent process with all inter-loop coupling effects embedded. In this way, the controller can focus on a sole objective of performance. A novel approach is developed to obtain the control parameters so that a point on the Nyquist curve is moved to a desired position for each equivalent process. Furthermore, guidelines to determine the destination point are provided. The extension to cross-coupled multivariable controllers is briefly addressed.

Chapter 6 is focused on the common practical implementation issues and problems associated with the use and installation of PID controllers. Problems arising from non-idealities in the control system, such as non-linearities in the various elements of the system and disturbances occurring at various points in the process, are highlighted. Possible practical solutions are suggested. Useful operational functions such as, set point weighting and auto-manual transfer for the PID controller, are also elaborated. The last section examines the issues arising from a digital implementation of the PID controller.

Finally, a sample of industrial PID controllers, their functions and special features are given in the appendix to the book. The appendix is intended to provide readers some practical insight of typical industrial controllers and it also serves to provide a log of the status of controllers just before the dawn of Year 2000.

This book is written with tremendous contributions from the following persons : Mr Jiang Xi, Mr Leu Fook Meng, Ms Raihana Ferdous, Mr Seet Hoe Luen, Mr Zhang Yu, Mr Zhang Yong and Dr Bi Qiang. The authors would like to express their gratitude and appreciation to them.

CONTENTS

CHAPTER 1

INTRODUCTION

1.1 Evolution of the PID Controller

Automatic control was first implemented some 60 years ago in the form of complex pneumatic mechanical gadgets, paving the way for a new and revolutionary method for producing the basic commodities of that time. The pasteurization process is one of the first application areas of automatic control (Babb, 1990). The process is simple - the temperature of a batch of milk is raised to 143 deg F and held there for 30 minutes to complete pasteurization. Around 1912, the dairy industry began logging and keeping records of the pasteurization process, due probably to the American government having more direct control over pasteurization. This was good news for thermometer manufacturer Taylor Instruments, based in Rochester, N.Y., which has not been doing too well with the sales of the mercury thermometers.

The reversal in fortune began with Iowa-based dairy equipment manufacturer Cherry-Burrel putting in an order with Taylor for 50 temperature recorders. Taylor engineers and salesmen recognized the opportunity and reacted quickly to capitalize on it. The original order in 1912 was the beginning of decades of domination of the dairy industry by Taylor. From their factory in Rochester, different customized temperature recorders were produced. In fact, Taylor's temperature recorders became the de facto standard of the dairy industry. Competitors such as Foxboro and Brown were virtually locked out of the race at that time. Even as late as 1940, dairy sales of $625,000 accounted for around 17% of Taylor's business.

Pasteurization technology changed in the mid-1930s. A new, nearly-continuous "flash" process heated the milk to a higher temperature of 160 deg. F for 15 seconds. If the temperature was not precisely maintained, a quick-acting diverting valve would activate to flush the milk back into the heating tank. The instrument was, in fact, directly and automatically controlling a part of the process. The pasteurization process was greatly accelerated, but essentially, it remained a batch operation, like other chemical and refining processes in the early part of the century.

The first real move towards continuous flow processing was probably taken around 1925 by Carbide & Carbon Chemicals. The company was experimenting with fractionation of natural gas and discovered new synthetic chemicals could be produced, such as ethyl alcohol and ethylene glycol. However, storing the natural gas was a more cumbersome affair than storing the raw crude oil, and so Carbide investigated ways for the delivery of the natural gas through their crackers more quickly, so as to alleviate the storage problem.

Taylor was approached to supply the recorders and a few controllers for the process. But the technology team did not succeed for a few reasons. First, Carbide process technology was kept highly confidential and little was revealed to the Taylor engineers. Secondly, there was a major technology barrier: the gas pipes required bigger and more tightly packed valves. It was hard to open and shut them, and Taylor's pneumatic actuator was less than effective. By 1933, when Taylor came up with a suitable valve positioner, it was already redundant as Carbide operators had become skilled in manual continuous-flow operation.

Meanwhile, deep in the heart of Texas, the business of oil refining was booming. The petroleum refiners were buying up all the crude the oil fields could produce. By 1929, they began to explore continuous flow as a means of increasing their competitive advantage. They built continuous-cracking furnaces to handle the job, but due to the increased speed of the operation, refiners had to rely on instrument control more heavily than ever before. They also had a great need for more accurate flow measurement.

For years, Foxboro had been a primary source of instrumentation for the oil fields. It was practically the only company with enough flow meter technology for the operations. Foxboro had a good opportunity to expand its operation into the refineries, and to install entire complexes with Foxboro instrument rigs. Taylor was basically out of competition of this nature as they did not make a single flow meter. Meanwhile, Foxboro's instruments and Stabilog controllers became the de facto standard in the oil, gas, and refining industries. The business was so good that by 1933, Brown introduced new instruments and joined the race. While Foxboro dominated, there was enough business opportunity for Brown to also have a significant share.

While Brown and Foxboro went after the booming refinery business, Taylor pursued a different path. It chose instead to concentrate on developing and improving their trademark controller. The first Taylor controller (Model 10R) was a non-indicating or fixed, proportional controller. It used either a pressure capsule for pressure control or a vapor temperature bulb for temperature control. The set point was changed mechanically by a cam which operated a poppet valve air relay. The first truly adjustable proportional controller by Taylor was the Model 56R which was made in 1933. It was called the Fulscope and was the first proportional controller with a proportional adjustable range.

Sensitivity (or proportional band) was adjustable by a knob from 1,000 psi/in. to about 2psi/in.

Foxboro came out with their Model 40 around the period 1934-1935 (Blickley, 1990). It was probably the first proportional plus reset recorder and controller. It was used principally in flow control in the petroleum industry. The operational principle was to dampen the mercury manometer so that the sensitivity or gain could be set high. The reset action was caused by spools of capiliary which were changed for different reset rates. Some of the related problems included slow response, overshoot on startup, and offset.

It was not too long after this, in 1935, that Taylor designed the "double response unit" which was a gadget that provided automatic reset on top of the proportional control from the Fulscope. It has feedback from the valve stem and later became the first valve positioner in the industry, in 1936. The double response unit was very complicated, had poor instructions, and was difficult to adjust, causing it to be nicknamed the "dubious response unit". When properly set, it would compensate for load changes, and it could be set to under-compensate or over-compensate. It provided stable control with no offset.

Taylor, around this time, was also working with the viscose rayon industry, trying to control the rayon shredder. The proportional Fulscope with the double response unit was found to be ineffective for this application. The research department further investigated the Fulscopes, and in the process, put a restriction in the feedback line to the capsule that made the follow-up in the bellows. This gave a strange "kicking" action to the output. The technique was tested on the rayon shredders and it gave perfect control on the temperature. The action was dubbed "Pre-Act" and was found to help the control in other difficult applications such as the refinery processes. The Pre-Act was the first derivative control and it was incorporated into the Model 56R and worked effectively on juice units in the sugar industry, but it was not too useful in other applications. The Model 100 Fulscope was subsequently designed in 1939 incorporating the automatic reset action provided by the double response unit.

Ken Tate was the head of the Taylor engineering department when the 100 series was being designed, and he was responsible for the ingenious parallelogram linkage changing the mechanical feedback to make a continuous adjustment of proportional response sensitivity. Bill Vogt, second in command, designed the reproducible needle valves for setting reset rate and pre-act time. This was when the very first proportional-plus-reset-plus-derivative control came integrated in one unit.

1.2 Components of the PID Controller

The PID control structure consists of three constituent components, the Proportional, Integral and Derivative part. This section will illustrate the functional principle of each of these components and explain their evolution from the simplest of all types of controller, namely the On/Off controller.

The On/Off controller is undoubtedly the most widely used type of control for both industrial and domestic service. It is the simplest and most intuitively akin to direct manual control. It has an output signal u which may be changed to either a maximum or a minimum value, depending on whether the process variable is greater or less than the set point. The control law is described by:

$$\begin{aligned} u &= u_{max}, \quad e > 0, \\ &= u_{min}, \quad e < 0, \end{aligned}$$

where e is the control error $e = r - y$ (Fig. 1.1). The minimum value of the control output is usually zero (off). It has been assumed that the process has a positive static gain, but On/Off control is equally applicable to one with a negative static gain as well, with a direct exchange of the switching conditions. The mechanism to generate On/Off control is usually a simple relay.

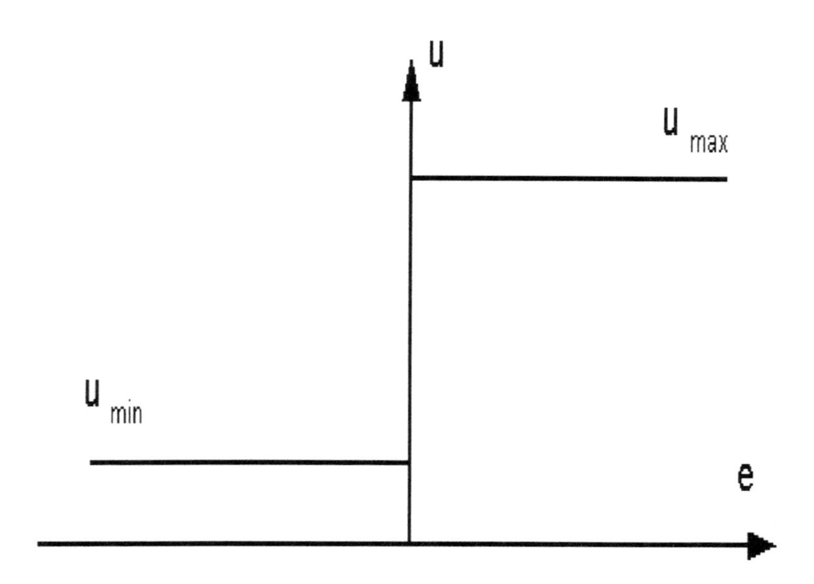

Fig. 1.1. ON/OFF control

One renowned and significant disadvantage of this controller is that it oscillates around a constant set point. This directly affects the process variable which will also oscillate around the desired value. If, for example, On/Off control is applied to controlling the level in a tank with the aid of an On/Off valve which can only open or shut completely, the On/Off controller therefore naturally open and shut the valve alternatively. While this control method is attractively simple and easy to implement, it causes rapid wear and tear on the moving parts in the actuating device (the valve in this level-control case) and the "ringing" or oscillation phenomenon may not be tolerable in certain cases. The solution in most On/Off controllers is to establish a dead zone or hysteresis of about 0.5% to 2.0% of the full range. This dead zone straddles the set point so that no control action takes place when the process variable lies within the dead zone.

1.2.1 The Proportional part

Apart from the use of a small deadzone to reduce the phenomenon of signal oscillations, one alternative way to alleviate the oscillation phenomenon associated with On/Off control is to use a small gain for the controller when the error e is small and conversely to use a large gain when the error is large. This can be achieved with a proportional or P controller. Proportional control is the basic continuous control mode. The control signal in a P controller is given by:

$$
\begin{aligned}
u &= u_{max}, \quad e > e_0, \\
&= u_0 + K_c e, \quad -e_0 < e < e_0, \\
&= u_{min}, \quad e < -e_0,
\end{aligned}
$$

where u_0 is the level of the control signal when we have no control error, and K_c is the proportional gain of the controller. K_c is also referred to as the proportional sensitivity of the controller. It indicates the change in the control signal per unit change in the error signal. It is indeed an amplification and represents a parameter which may be adjusted by the operator. The P controller can also be described graphically as shown in Fig. 1.2.

There are several additional names for proportional control, such as correspondence control and modulating control. While the signal oscillations with On/Off control may be quenched, a new problem with proportional control now arises instead. With pure P control, it is possible and typical that steady state errors will occur. In other words, after the transients have died down, there may remain a deviation between the set point and the process variable.

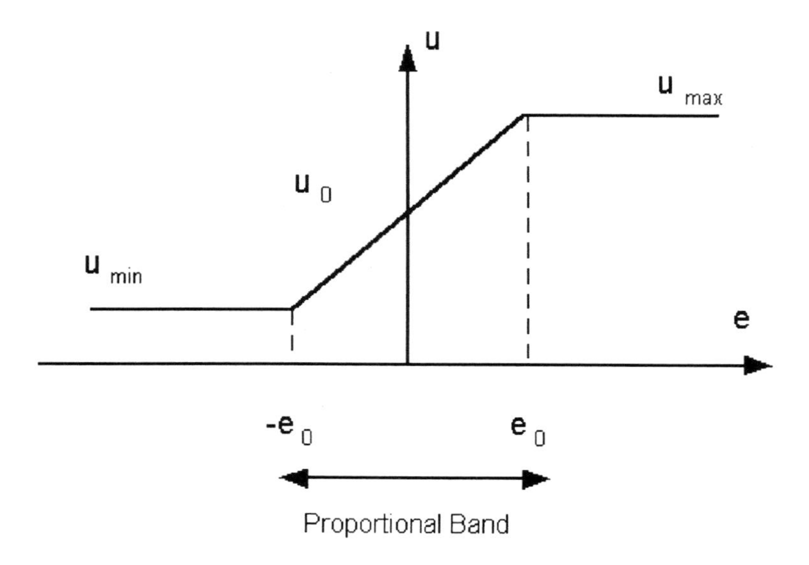

Fig. 1.2. P control

This phenomenon may be easily observed from the proportional relationship between the control signal and a small control error:

$$u = u_0 + K_c e.$$

This means that the control error is given by:

$$e = \frac{u - u_0}{K_c}.$$

In the steady state, the control error $e_{ss} = 0$ if and only if at least one of the two following conditions is true:

- K_c is infinitely large,
- $u_{ss} = u_o$.

Note that e_{ss} and u_{ss} denote the steady state error and control signal respectively. The first condition implies an infinite gain for the controller which effectively reverts P control back to On/Off control with the associated problems P control is set to resolve. The second condition cannot generally be satisfied for all set points r. Even if u_0 can be adjusted relative to the set point, it is still necessary to know at least the process static gain before the adjustment can be done.

Thus, for P control, the highest control gain may be used corresponding to an acceptable level of closed-loop stability to reduce the steady state error. In addition, the largest steady state error which can occur can be minimized if u_0 is selected to be right in the middle of the operating range of the control signal. In most controllers, therefore, u_0 is chosen to be 50%.

1.2.2 The Integral part

To resolve the steady state error problem with P control without using excessive controller gain, the integral or reset action should be introduced. The integral action eliminates the problem of any remaining steady state error. The I part is able to find the correct value for u_0 automatically in response to any set point without having to know the process static gain. Integral control action usually is combined with proportional control action, although it is possible but less common that integral action is used by itself. The combination is referred to as PI control. The combination is favorable in that some of the advantages of both types of control action are made available.

The control signal in a PI controller is given by:

$$u = K_c \left(\frac{1}{T_i} \int edt + e \right),$$

where T_i is the integral time of the controller. The constant level of u_0 found in the P controller has thus been replaced by the integral:

$$u_0 = \frac{K_c}{T_i} \int edt.$$

The integral of the control error is effectively proportional to the area under the curve between the process variable and the set point (Fig. 1.3).

To better understand the steady state error elimination capability of the PI controller, assume that the closed-loop system is stable and that a steady state control error existed despite having a PI controller which implies $e_{ss} \neq 0$. The integrator will continuously accumulate the error signal at the input, and thus the control signal u will be either rising or falling, depending on whether the error is positive or negative. The P part has a constant value corresponding to $K_c e_{ss}$ and thus will not affect the analysis. If the control signal rises (or falls), the process variable will also rise or fall. This in turn means that the error $e = r - y$ cannot be constant in steady state and thus contradicts the assumption that the error is stationary. Thus, it is not possible to have a non-zero steady state error when the controller has an integral part and the closed-loop is stable.

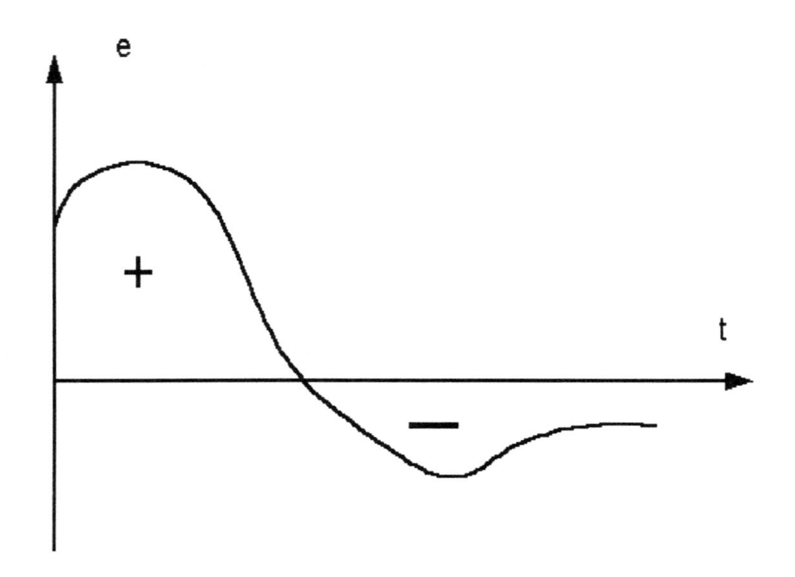

Fig. 1.3. Integral control derived from the area under the control error curve

A PI controller thus solves the problem of the remaining stationary error and the problem of oscillation associated with On/Off control. The PI controller is therefore an efficient controller without any significant faults and is often sufficient when the control requirements are low to modest.

1.2.3 The Derivative part

Both the P and the I components of a PI controller operate on past control errors, and do not attempt to predict future control errors. This characteristic limits the achievable performance of the PI controller. The problem is more clearly illustrated in Fig. 1.4. The two curves in this figure show the time graph of the control error for two different processes. At time t, the P part, being proportional to the control error, is the same for both cases. Assuming that the I part, being proportional to the area under the two control error curves, are also equal in the two cases. This means that a PI controller gives exactly the same control signal at time t for the two processes. However, it is clear that there is a great difference between the two cases if the rate of change in the control error is considered. For Response I, the control error changes rapidly and here the controller should reduce its output to avoid an overshoot in the process variable occurring in the future. For Response II, the control error changes sluggishly and the controller should react strongly in order to reduce the error more rapidly. Derivative or rate control indeed carries out just this type of compensation.

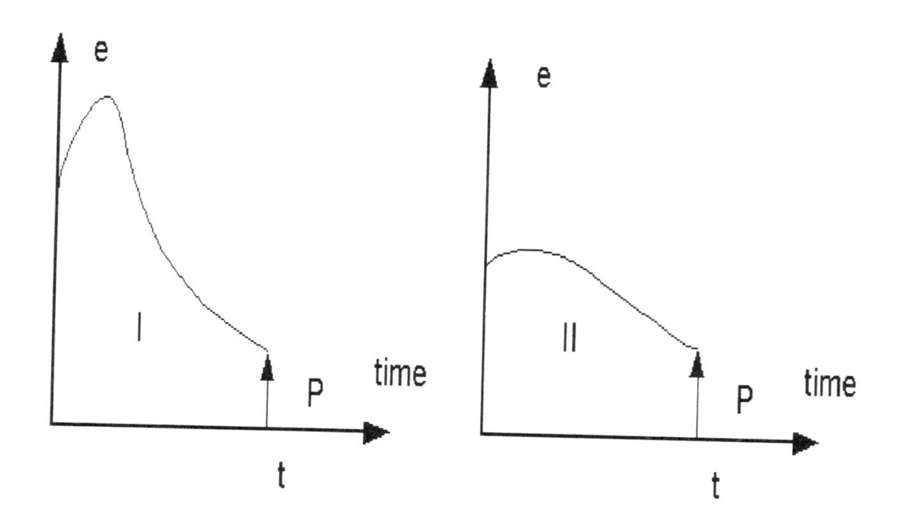

Fig. 1.4. Illustration of the need for derivative control

While it is possible theoretically to have a control action based solely on the rate of change of the error signal e, it is not practical since if the error is large but unchanging, the derivative controller output would be zero. Thus, derivative control is usually combined with at least a proportional control.

The D part of the PID controller is proportional to the change in the error, in other words its derivative (Fig. 1.5). The D part is proportional to the predicted error at time $t+T_d$, where T_d is the derivative time of the controller. The control law for the PID controller is:

$$u = K_c \left(e + \frac{1}{T_i} \int edt + T_d \frac{de}{dt} \right).$$

In practice, the derivative is not usually taken of the control error, but only of the process variable. This is because set point r is normally constant with abrupt changes. It will thus normally not contribute to the derivative term. Moreover, the term dr/dt will change drastically when the set-point is changed. For this reason, it is common practice to apply the derivative action only to the process output.

The control structure is then:

$$u = K_c \left(e + \frac{1}{T_i} \int edt - T_d \frac{dy}{dt} \right).$$

There is generally no noticeable difference in control performance between these options; stability or the ability to deal with disturbances or load changes are unaffected, and derivative on the process variable is normally the preferred choice. The only clear occasion when true derivative on error is advantageous is where the process variable is required to track a continually changing set point as in military gunnery control.

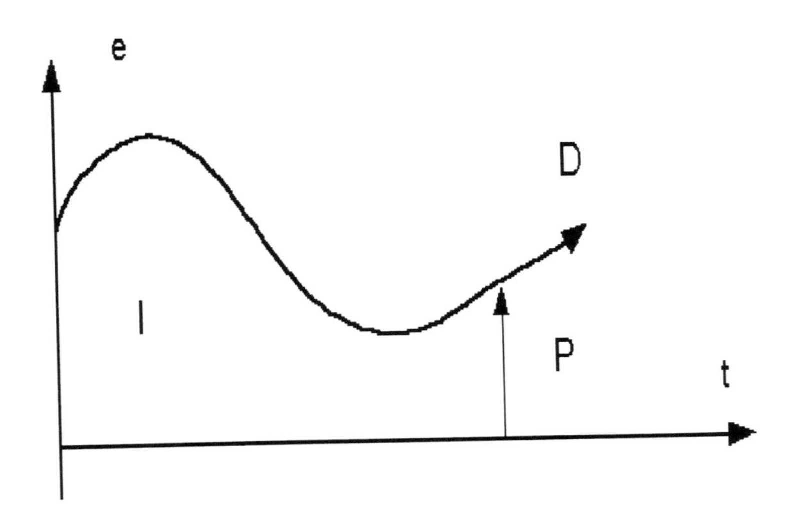

Fig. 1.5. Components of PID controller

The derivative action may also result in difficulties if high frequency measurement noise is present. In practice, these difficulties are normally resolved using additional derivative filtering techniques (Chapter 6).

1.3 Choice of Controller Type

The PID controller thus consists of three components, each with its own distinctive function to fulfill certain control objectives. In actual applications, different permutations of the P, I and D components may be used depending mainly on the process and the control requirements. This section serves to provide some guidelines on the choice of the more common controller members of PID family and its precedent, the On/Off controller.

1.3.1 On/Off controller

The simplest type of controller, the On/Off controller, has the well-known disadvantage that it causes oscillations in the process variable. However, it also has great advantages which account for the persistent and continued application of this controller. It is very simple, rugged and therefore come tagged with a low cost. It does not require the adjustment of any controller parameters. Therefore, in circumstances where oscillations are not too much of a problem and inexpensive design is important, On/Off controller should be considered based on the TSTF engineering principle (Try Simple Things First). This class of controllers is still predominant in simple domestic appliances such as water heaters, air-conditioners, ovens, refrigerators, stoves, irons etc.

1.3.2 P controller

P control is close in simplicity to On/Off control with only one additional tuning parameter in the proportional gain. The major factor affecting its application is the steady-state errors typically associated with pure P control. For low-order processes (e.g., single-capacitive and integrating processes) where high control gain can be used without endangering stability, P control may be used with minimum steady state offset between the set point and process variable. In these cases, there is no necessity for an integral part in these control examples, if the remaining control error is acceptable. In surge tanks, for example, the exact level control is not important, but rather the operating level must remain within acceptable limits, such as the working range of the level transducer. Pure P control is also sufficient in the inner loop of cascade control systems *(configurations with one secondary loop within a primary loop)* (Luyben 1990) where asymptotic set point regulation is left to the outer loop controller.

As discussed, if the static gain of the process is known and constant, then u_0 may be adjusted exactly to the set point and very good performance in terms of steady state errors can be achieved.

1.3.3 PD controller

There are systems with inherent integrator dynamics. One prominent example is a servo positioning mechanical system. A well-insulated thermal process with a large time constant has the same integrating effect, since with no parasitic heat losses, all the energy supplied goes to raise the temperature of the process. With these types of process, no further integration is necessary of the controller. On the other hand, pure P control is usually insufficient for processes with true or equivalent integration action due to long time constants. These processes has extensive lag around the loop. By adding a D part to the P controller, phase lead is added which is appealing in speeding up the dynamical response. The derivative component predicts future errors to react ahead of time to avoid excessive overshoots or undershoots to set point changes. This derivative part allows energy input to be varied appropriately in time. In thermal processes with large time constants and of the second or third order, it may not be sufficient even to use the derivative part and higher derivatives of the temperature may also have to be used. This means that we are not just studying the changes in temperature via the derivative part, but also the acceleration of the temperature changes. One associated problem with PD control is its sensitivity to noise because it has a relatively large gain at high frequencies. Thermal processes have relatively noise-free process variables which is the main reason why PD controllers have remained effective with these processes.

1.3.4 PI controller

This is the most common and popular controller structure in the PID family. It is generally applicable to many processes encountered without any significant problems such as oscillation or steady state control error. The D part is unnecessary if there are no high demands on the speed of the control loop. It is also not required either if the process is of a high-order in inertia with large phase shifts (multiple capacitive processes). These processes do not respond immediately to a change in the control signal as can be seen clearly in Fig. 1.6 which shows the step response for such a typical process. The effect is rather similar to the presence of process dead-time which is why high-order systems are sometimes modeled with a lower-order model with an appropriate dead-time. Although for these kinds of processes, there is the greatest need for process prediction, yet such prediction cannot be effectively obtained with the D component of the PID family. Because of the dead-time, true or equivalent, there will be a time delay before the effects of any control action can be seen on the process variable. The D component, if present, will attribute the lack of responsiveness on the part of the process to insufficient control effort and increase the control correspondingly. The cumulated effect will subsequently cause very large control errors or even instability in the closed-loop performance. Thus, for these processes, the D part is best disconnected and PI is more appropriate. Furthermore, the use of D component, despite filtering, results in amplification of noise which then appears in the control signal. More effective ways of dealing with time-delay processes are the use of dead-time compensation such as the Smith predictor controllers.

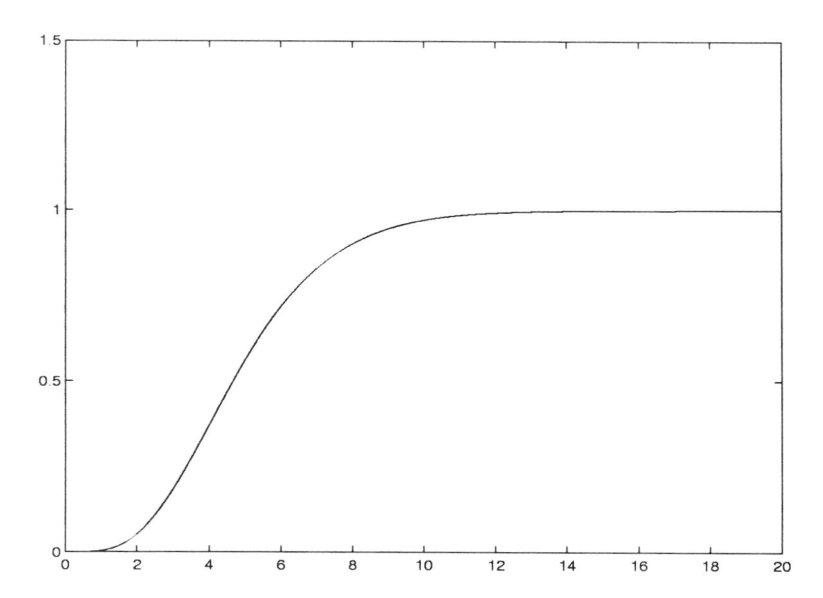

Fig. 1.6. Step response characteristic associated with time-delay or high-order processes

Another case where PI is sufficient is when the process has a small phase shift (low-order, single-capacitive processes). Here with little or no inertia in the process, the D part becomes unnecessary and it does not give any great improvement to the control but may become a cause of trouble due to the inherent noise amplification property.

1.3.5 PID controller

PID controller is the most sophisticated continuous controller among those discussed. Following from earlier discussions, the D part is best used on processes without long dead-times but with larger phase shifts (e.g., double or triple-capacitive processes). In these cases, the D part gives better controller performance than if it was not present. It allows larger gains for both the P and the I part. PID thus may give rapid response and exhibits no offset, but it is difficult to tune - now there are three knobs to adjust. As a result, it is used only in a very small number of applications, and it often requires extensive and continuing adjustment to keep it properly tuned. It does, however, provide very good control when good parameters are implemented.

1.4 Nomenclature of the PID Controller

The nomenclature for various parameters of the PID controller changes from one product to another. In many controllers, the gain-adjusting mechanism on many industrial controllers is not expressed in terms of the proportional gain, but in terms of the proportional band PB . Proportional band is defined as the span of values of the input which corresponds to a full or complete change in the output. The relationship of PB to K_c is:

$$PB = \frac{100}{K_c}[\%].$$

The gain $K_c = 1$ thus represents a proportional band of $PB = 100\%$. The relationship is also shown diagrammatically in Fig. 1.2. As a matter of practice, *wide bands* (high percentages of PB) corresponds to a less sensitive response and *narrow bands* (low percentages) correspond to a more sensitive response.

Manufacturers sometimes also uses "reset" to refer to integral action and "pre-act" for derivative action. One is the reciprocal of the other.

$$T_i = 1/\text{reset},$$
$$T_d = \text{pre} - \text{act}.$$

1.5 Structures of the PID Controller

PID controllers come in various structures, depending on the manufacturer. One main reason for the non-standard structure is due to the transition of the controllers from pneumatic-based implementation to microprocessor-based implementation of the present age and the corresponding change in the structural requirements. Many manufacturers also stick to past conventional structures due to the large base of their users already familiar with tuning the controllers in the old structures.

The different structures of PID controller have direct implications on controller tuning. If the controllers are tuned by trial and error methods, the implications are not large. The rules of thumb for the adjustment of PID parameters is usually valid for all designs. For a systematic tuning of a PID controller based on structural assumptions such as many of the methods to be reported in Chapters 2 and 3, it is important to know the structure of the controller. Structural difference also becomes important when one product line is replaced with another. The old controller gains cannot be directly ported over to the new controllers. This section serves to illustrate the main structures of today PID controllers - the parallel and the series type.

1.5.1 Parallel type

The parallel type was the one described earlier, with the control law:

$$u = K_{cp}\left(e + \frac{1}{T_{ip}}\int edt + T_{dp}\frac{de}{dt}\right).$$

In many literature, this is also known as the ideal type. For the sake of simplicity, the discussion here will be based on this basic structure without considering other minor structural modifications for derivative filtering etc. These issues will be addressed in details in Chapter 6.

The parallel type can be shown diagrammatically as in Fig. 1.7. It is the one normally described in text books and the more general compared to the series type of PID controller. The main characteristic for the parallel type is that the P, I and D parts are separated and connected in parallel. Until recently, this type is not commonly seen in industrial process controllers. The main reason is that this form of PID control was difficult to design using pneumatic controllers due to the requirement for more analog amplifiers which were very expensive at that time. Traditionally, this type has not been adopted, even though it is just as easy to implement as the other designs using modern microcomputer-based controllers. However, with wide proliferation in microprocessor technology, the parallel form of PID control is poised to become more and more common in modern control systems.

1.5.2 Series type

The series type is still the most common form of industrial process controllers. It may be described by the following equations:

$$e_1 = e + T_{ds}\frac{de}{dt},$$

$$u = K_{cs}\left(e_1 + \frac{1}{T_{is}}\int e_1 dt\right).$$

This structure can be described diagrammatically as in Fig. 1.8. In the series type, the I and D parts are not independent as in the parallel type, but the I part works on both the error and the derivative part. The controller may be regarded as a series connection of a PI and a PD controller. The series and parallel types of controller are therefore often also known as interacting and non-interacting types respectively. By using the interacting form, a three-term controller can be made with only one amplifier. Thus, pneumatic

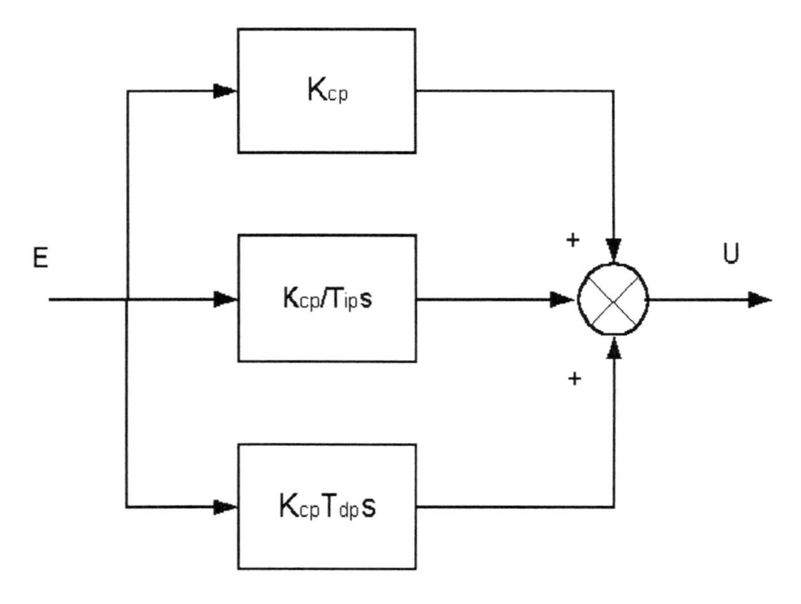

Fig. 1.7. Structure of parallel type PID controller

controllers and early electronic controllers often used the interacting form to save on amplifiers which were expensive at the time. Some manufacturers deliberately use the interacting form in their digital algorithms in order to keep the tuning process similar to the tuning of electronic and pneumatic controllers.

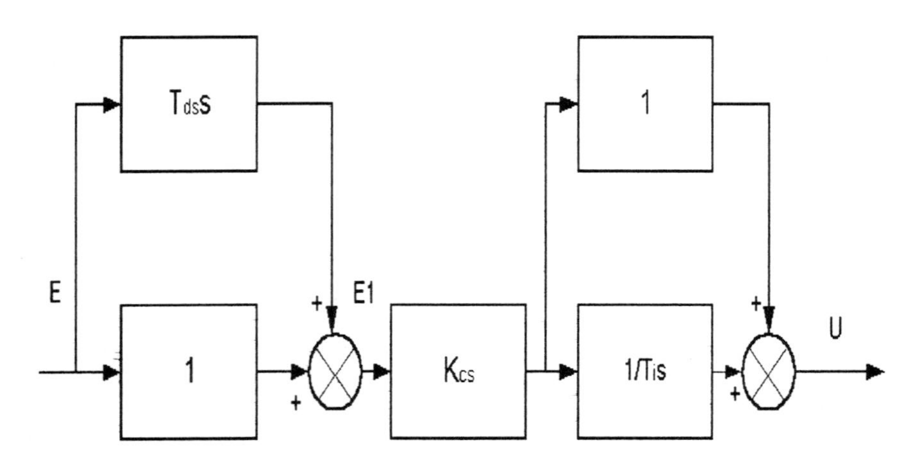

Fig. 1.8. Structure of series type PID controller

Other PID controller structures have been proposed in the literature, each suitable for a specific requirement. For interests, the readers may refer to (Frank, 1968), (Eitelberg, 1987), (Hippe et. Al., 1987), (Mantz and Tacconi, 1989).

1.5.3 Relationship between Parallel and Series types

The relationships between the PID parameters in these two controller structures can be easily obtained. Given the parameters for the series type, the corresponding parameters for the parallel type can always be obtained and they are given by:

$$
\begin{aligned}
K_{cp} &= K_{cs}\frac{T_{is}+T_{ds}}{T_{is}}, \\
T_{ip} &= T_{is}+T_{ds}, \\
T_{dp} &= \frac{T_{is}T_{ds}}{T_{is}+T_{ds}}.
\end{aligned}
$$

On the other hand, given the parameters for a controller of the parallel type, it is not always possible to obtain the corresponding parameters for the series type. The condition under which this is possible is:

$$T_{ip} > 4T_{dp}.$$

This means that the parallel type is more general than the series one. The relationship between the parameters will then be:

$$
\begin{aligned}
K_{cs} &= \frac{K_{cp}}{2}\left(1+\sqrt{1-\frac{4T_{dp}}{T_{ip}}}\right), \\
T_{is} &= \frac{T_{ip}}{2}\left(1+\sqrt{1-\frac{4T_{dp}}{T_{ip}}}\right), \\
T_{ds} &= \frac{T_{ip}}{2}\left(1-\sqrt{1-\frac{4T_{dp}}{T_{ip}}}\right).
\end{aligned}
$$

It may be noted that the series and parallel types only differ for a PID controller, i.e., when all three parts in the controller are used. For a P, a PI or a PID controller, the two types are identical.

The different controller structures become a problem if the make of controller is changed. In this case, the tuning procedure have to be carried out again, or the controller parameters recalculated according to the above relationships.

Finally, it should be added that there are other factors which can cause one set of controller parameters which give good control for one product to give worse control on another product. Such factors include filtering and special logic on changes of set point, such as set point weighting to be examined in Chapter 6.

1.5.4 Incremental type

Diaphragm operated actuators can be arranged to fail open or shut by reversing the relative positions of the drive pressure and return spring. In some applications, a valve will be required to hold its last position in the event of failure. One way to achieve this is with a motorized actuator, where a motor drives the valve via a screw thread.

Such an actuator inherently holds its last position, but the position is now the integral of the controller output. An integrator introduces $\pi/2$ phase lag and gain which falls off with increasing frequency. A motorized valve is a destabilizing influence when used with conventional PID controllers, parallel or series.

Incremental PID controllers are designed for use with motorized valves and similar integrating devices. They have the control algorithm (illustrated for the parallel type, but equally applicable to the series type):

$$u = \frac{d}{dt}\left(K_c\left(e + \frac{1}{t_i}\int edt + T_d\frac{de}{dt}\right)\right).$$

which is the time derivative of the normal control algorithm. Incremental controllers are sometimes called boundless controllers or velocity controllers because the controller output specifies the actuator rate of change (i.e., velocity) rather than actual position. Incremental controllers is thus immune against integrator windup (Chapter 6) since the controller output can be directly limited, but it is often undesirable to keep driving a motorized valve once the end of travel is reached. End of travel limits are often incorporated in motorized valves to prevent jamming. The controller also has no real "idea" of the valve true position, and hence cannot give valve position indication. If end of travel signals are available, a valve model can be incorporated into the controller to integrate the controller output to give a nominal valve position. This model would be corrected whenever an end of travel limit is reached.

CHAPTER 2
CLASSICAL DESIGNS

2.1 Introduction

In Chapter 1, the PID control structures and components have been examined. In this chapter, simple methods for tuning the controller, i.e., selection of K_c, T_i and T_d will be provided. These include a heuristical approach based on trial and error, and several classical systematic methods which involve conducting an experiment, determining a process model and using known tuning methods to determine the control gains instead of a trial and error basis. Systematic methods are worthwhile and necessary for certain cases:

1. Important loops where control performance is critical and important.
2. Difficult loops where trial and error methods, and the usual rules of thumb are not applicable or effective.
3. Slow loops which will take far too long time to test each new setting for trial and error methods to be feasible, in which case it is often worth calculating a setting using systematic approaches.

In this chapter, simple classical methods which give the controller parameters directly from the step or frequency responses are illustrated which do not require any long calculations, but where the controller parameters can be obtained rapidly by hand or with a calculator. Although in certain cases, tables or formulae are given specifically for the parallel type of PID controller, they may be easily converted into equivalent ones for the series type according to Section 1.5.3.

2.2 Design Objectives - Speed Versus Stability

The design objectives or the way the controller is expected to control naturally depends on the application in point. The general dilemma faced in controller tuning is the compromise between the desire for speed versus the desire for stability . Fast control is usually accompanied by poor stability and oscillations. A step change in the set point may well result in a severe overshoot . On the other hand, very stable control without overshoot is usually achieved at the expense of a more sluggish response to set point changes and load disturbances. The solution to this compromise may also depend on the type of disturbance present. In Chapter 6, the practical issue of disturbances affecting controller tuning and control performance will be more closely examined. The robustness of the controller is also related to the demand for speed. Robustness means the sensitivity to variations in the process dynamics. Controllers which are tuned to give fast control are usually more sensitive to variations in the process than controllers which are more conservative in their settings.

The most common criterion behind the principles of many classical and systematic way of tuning the controller is to adjust the controller so that the system response curve has an amplitude decay ratio of one-quarter. A decay ratio of one-quarter means that the ratio of the overshoot of the first peak in the process response curve to the overshoot of the second peak is four to one. This is illustrated in Fig. 2.1. This is a good tuning criterion if we are primarily interested in dealing with load disturbances. On the other hand, these methods often produce rough control in cases of set point changes.

Basically, there is no direct mathematical justification for requiring a decay ratio of one-quarter, but it represents a compromise between a rapid initial response and a fast response to disturbances. In many cases, this criterion is not sufficient to specify a unique combination of controller settings, i.e., in two-mode or three-mode controllers, there are an infinite number of settings which will yield a decay ratio of one-quarter, each with a different period. This illustrates the problem of defining what constitutes good control.

In certain cases, it is important to tune the controller such that there is no overshoot; in certain cases, a slow and smooth response is desired; other cases may warrant fast response and significant oscillations are no problem. The point is - the definition of good control depends on specific requirements.

2.3 Trial and Error Method

The PID controller has three adjustable parameters, the proportional gain K_c, the integral time T_i and the derivative time T_d. In certain cases, the operator can also set a filter time constant in order to be able to use the derivative part (see *Derivative Filtering* in Chapter 6).

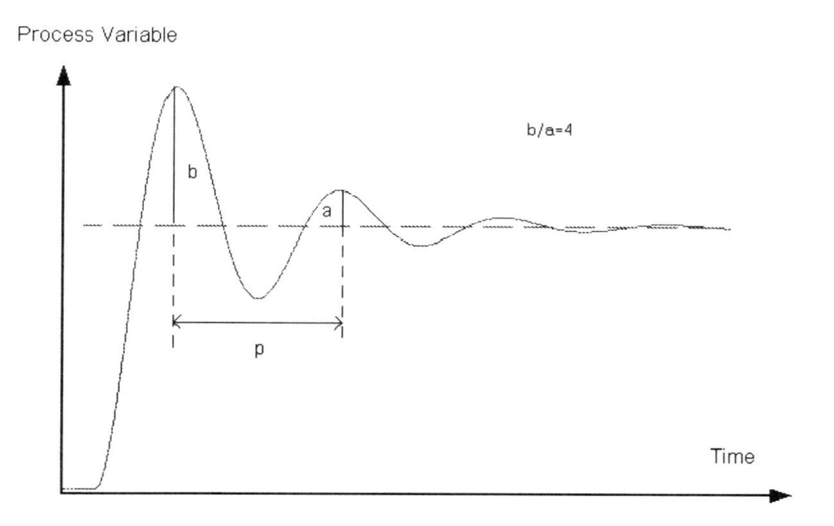

Fig. 2.1. One-quarter decay ratio criterion

In this section, the general and typical effect from the individual parameters on the performance in terms of response speed and closed-loop stability will be elaborated. It is naturally necessary and important to know this "cause and effect" when the controller is to be tuned without the help of systematic methods, i.e., when tuning a controller purely by trial and error. Even when the systematic methods are used, it is still important to know how the different controller parameters affect the control, because these methods do not always result in the desired control performance and subsequent fine tuning and manual adjustments are necessary.

The selection of the controller parameters means finding a compromise between the requirement for fast control and the need for stable control. Table 2.1 shows how stability and speed change when the parameters are changed. Note that the table only contains rules of thumb and there are exceptions. For example, an increased gain often results in more stable control for low-order processes such as those involving liquid level control.

When a PID controller is tuned manually, it is usual to tune the parameters in the order P, I, D, i.e., starting with K_c, then setting T_i followed by T_d. Initially, the I and D parts are effectively disconnected by setting T_i very

Table 2.1. Effects of PID parameters on speed and stability

	Speed	Stablity
K increases	increases	reduces
T_i increase	reduces	increases
T_d increase	increases	increases

high (if the integral part cannot be switched off altogether), and $T_d = 0$. After K_c has been adjusted according to the rules-of-the-thumb in Table 2.1 such that the performance approaches the desired one, T_i is reduced to a suitable setting for the integral part. Referring to Table 2.1, this will result in a reduction in stability, which in turn means that the gain K_c has to be reduced. One significant exception to this is in liquid flow control. Liquid flow control loops are very fast and quite often tend to be very noisy. As a result, integral action is often added to the feedback controller in liquid flow control loops to provide a dampening or filering action for the loop. The advantage of eliminating offset is still present in these cases, but it is not the principal motivating factor.

When the PI controller exhibits satisfactory performance, the adjustment for T_d may begin. Increase in T_d will normally result in an improvement in stability, which in turn means that the gain K_c may be further increased and T_i further reduced.

The derivative part, according to the table, produces both faster and more stable control when T_d is increased. This is only true up to a certain limit, and if the signal is sufficiently free of noise. Raising T_d above this limit will result in reduced stability in control. As mentioned previously, the function of the derivative part is to estimate the change in the control a time T_d ahead. This estimation will naturally be poor for large values of T_d. The reasoning above has also been made without taking into account noise or other disturbances. The noise is amplified to a greater extent when T_d increases. It is thus often the noise which sets the upper limit for the magnitude of T_d. In some cases, the noise level can be so high, despite filtering, that one may be unwilling to use the derivative part because it gives such a bad control signal. A PI controller would then be preferred with its smoother control signal, even at the cost of poorer control performance.

2.4 The Ziegler-Nichols Methods

Until 1940, the tuning of process controllers was still more of an art conducted by ad-hoc methods on controllers that were a hodge-podge of techniques or add-on components that defied any logical rules that could be universally applied.

When two engineers at Taylor Instrument Company tried tuning the Fulscope controller, they discovered that tuning two parameters was already difficult, not to mention three for the full PID controller (Babb, 1990). The engineers, John Ziegler and Nathaniel Nichols, decided to work on the tuning problem. By 1941, they had a relatively straightforward method for tuning the PID controller, and when they decided to document the work, they changed the whole control industry. John Ziegler was the practical one of the pair with a lot of experience in process applications, and who performed all the simulator tests that led to the methods they were seeking. Nathaniel Nichols was the mathematician and who reduced all of the mathematics to a few simple relationships that could be understood by technicians and operators. The results were the now famous "Ziegler-Nichols" methods of tuning controllers - methods that survived the slings and arrows of its early detractors, withstood the test of time, and works just as well as many of the later more sophisticated optimizing forms on a great majority of process applications.

2.4.1 The step response method

The Ziegler-Nichols step response method is based on a step response of the uncontrolled process. A step input is made in the control signal to the process and the response is logged for analysis (Fig. 2.2).

Simple process characteristics are determined from the process in the following way:

1. Locate the point on the process response curve where the slope is greatest, and draw the tangent to the curve through this point.
2. Find the points at which this tangent cuts the two lines which give the stationary values of the process variable before and after the step disturbance respectively.
3. Reading off the times for these two points, the dead-time L and the dominant time constant T for the process can be estimated. The dead-time is defined as that time it took from when the step disturbance was made until the process signal began to react. With the Ziegler-Nichols method for determining dead-time, the estimated dead-time is often longer than

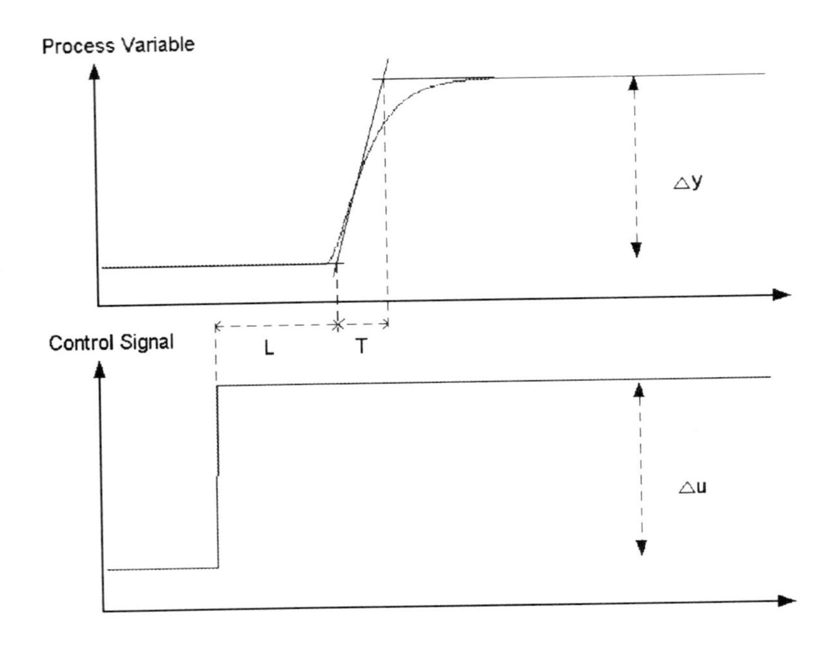

Fig. 2.2. Step response method

it is in real life. This is correct, and it is due to the very simple model of the process which is used since high-order dynamics also appear in the step response as an additional dead-time. In other words, the dead-time L and the time constant T have to describe a process which may comprise one dead-time and several time constants. This is approximated with a slightly longer dead-time and a dominant time constant.

4. The static gain K_p of the process is estimated by taking the ratio between the process variable change and the control signal change:

$$K_p = \frac{\Delta y}{\Delta u}.$$

Based on these three parameters L, T and K_p determined from the step experiment, the Ziegler-Nichols step response method can now provides the controller parameters (Table 2.2). In order to simplify the table, the relationship between the dead-time and the time constant has been designated by Θ, where

$$\Theta = \frac{L}{T}.$$

The constant Θ is known as the normalized dead-time. The Ziegler-Nichols table was originally drawn up for the PID controller in its parallel design type. The table is also filled with the series type using the calculations given in Section 1.5.3. Note again that it is only when all three parts of the controller (P,I and D) are used that there is any difference between the parallel and the series types.

Table 2.2. PID parameters according to Ziegler-Nichols step response method

Controller	K	T_i	T_d
P	$\frac{1}{K_p\Theta}$	-	-
PI	$\frac{0.9}{K_p\Theta}$	$3L$	-
PID (parallel)	$\frac{1.2}{K_p\Theta}$	$2L$	$L/2$
PID (series)	$\frac{0.6}{K_p\Theta}$	L	L

In the table, the controller gain is inversely proportional to the static gain of the process and the normalized dead-time. This is natural and intuitive. If the process has a high gain, the controller should compensate for this by having a low gain, and vice versa. If the normalized dead-time in the process is long, the process is difficult to control and a relatively low controller gain should be used under these circumstances.

Both the integral and derivative times are proportional to the dead-time of the process. This is also sensible, because the time parameters of the controller have to lie in the same range as the process time scale.

The main advantage of the Ziegler-Nichols step response method is in its simplicity as only a step experiment is required. The disadvantage is that the method is relatively sensitive to load disturbances and other disturbances in the frequency range of interest during the experimental phase and a large step input may be necessary to obtain a good signal-to-noise ratio. On the other hand, use of a large input is restricted by the non-linear modes of the process as well as important safety operational limits.

2.4.2 The frequency response method

In the Ziegler-Nichols frequency response method, the frequency response of the uncontrolled process is used to determine the controller parameters. In particular, the frequency response at the ultimate frequency, the frequency for which the phase lag of the process is $-\pi$. Ziegler and Nichols devised a method to obtain the ultimate frequency of the process in point.

Unlike the step response method where open-loop experiment is done, the Ziegler-Nichols frequency method is based on using the controller connected as a proportional controller. The experiment is carried out in the following way:

1. Put the controller into the automatic control mode with the I and D parts disconnected.

2. Increase the controller gain until the stability limit is reached, i.e., until the loop goes into self-oscillation. Measure the time period of the oscillation T_π.

3. From the oscillation period T_π and the gain of the P controller when the oscillation starts, k_π, the controller parameters can then be calculated according to Table 2.3. The period T_π and the gain k_π are known as the critical period and the critical gain respectively. Here, the controller parameters are given for both parallel and the series types of the PID controller. The advantages of this method are that it is very easy to generate the signal. The main disadvantage of the method is that the experiment operates on the verge of instability. Furthermore, the amplitude of the process variable signal may be so great that the experiment can often not be carried out for cost or safety considerations.

Table 2.3. PID parameters according to Ziegler-Nichols frequency response method

Controller	K	T_i	T_d
P	$0.5K_\pi$	-	-
PI	$0.4K_\pi$	$0.8T_\pi$	-
PID (parallel)	$0.4K_\pi$	$0.5T_\pi$	$0.125T_\pi$
PID (series)	$0.3K_\pi$	$0.15T_\pi$	$0.25T_\pi$

The method can be explained by studying the Nyquist diagram and the Nyquist criteria. The connection of a P controller and an increase in controller gain result in all the points on the Nyquist curve moving radially outwards from the origin. When K_c is increased to the extent that the Nyquist curve passes through the point -1, the stability limit is reached and self-oscillation is obtained. The dynamic gain at the ultimate frequency can then be calculated from:

$$k_\pi \left| G\left(j\omega_\pi = \frac{2\pi}{T_\pi} \right) \right| = 1.$$

The Ziegler-Nichols methods, both the step response and the frequency methods, are designed to give quarter amplitude damping.

Neither are the methods especially good for processes with long dead-times. Here, among the classical methods, the Cohen-Coon method (Section 2.6) is relatively better. As mentioned previously, it is best in this case to use just a PI controller. However, this should have a lower gain and a shorter integration time than those given by the Ziegler-Nichols methods. A good rule of thumb is:

$$K_c = \frac{k_\pi}{4},$$
$$T_i = \frac{T_\pi}{4}.$$

This tuning rule means that the controller works less with the proportional part and more with the integral part.

A similar version of the Ziegler-Nichols frequency response method, for PID controllers, is provided by David W. St. Clair formerly from Dupont, with extensive practical practice in instrumentation for tighter damping, compared to the quarter-damping-based tuning tules.

1. Increase gain till the loop maintains a small sustained cycle and note the oscillation period T_π.
2. Set $K_c = \frac{k_\pi}{2}$.
3. Set $T_i = T_\pi$.
4. Set $T_d = \frac{T_\pi}{8}$.

Interpretation of frequency response method. The Ziegler-Nichols frequency response method may be interpreted in terms of moving points in the Nyquist plot. The method begins with determination of the point $(-1/k_\pi, 0)$ where the Nyquist curve of the open-loop system intersects the negative real axis. With PI or PID control, it is possible to move a given point on the Nyquist curve to an arbitrary position in the complex plane. By changing the gain, it is possible to move the Nyquist curve in the direction of $G_p(j\omega)$, i.e., radially from the origin. Point A may be moved in the orthogonal direction by changing integral or derivative gain. It is thus possible to move a specified point to an arbitrary position, an idea that can be used to develop design methods (Fig. 2.3).

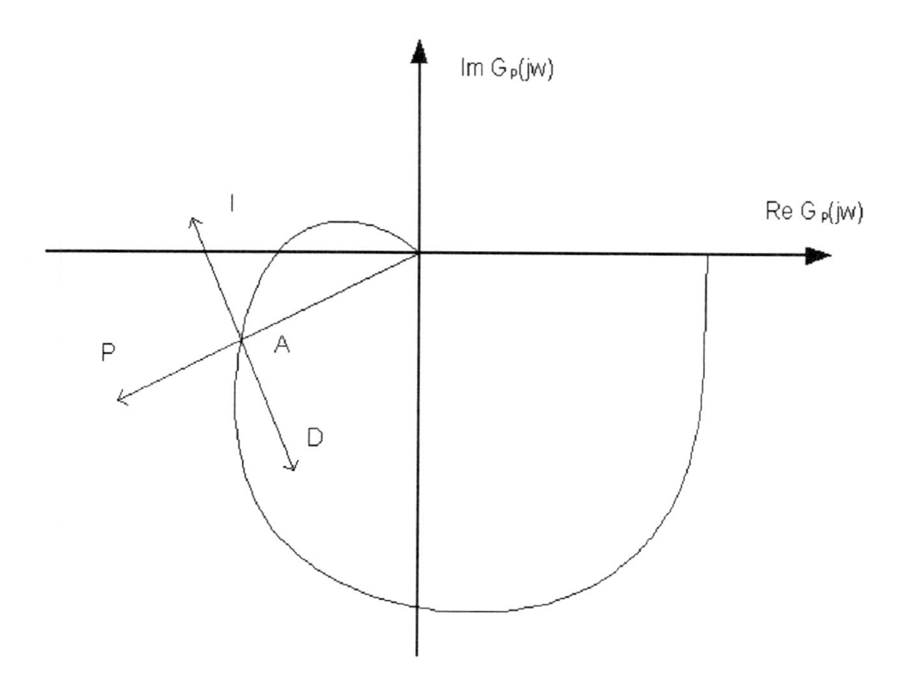

Fig. 2.3. Interpretation of the Ziegler-Nichols frequency response method

Let ω be the frequency that corresponds to point A. The frequency response of the PID controller at ω is

$$G_c(j\omega) = K_c\left(1 + \frac{1}{j\omega T_i} + j\omega T_d\right) = r_A e^{j\phi_A}.$$

With positive control parameters, the angle ϕ_A is thus restricted to the range $-\pi/2 \leq \phi_A \leq \pi/2$ where $\phi_A = -\pi/2$ corresponds to pure integral control and $\phi_A = \pi/2$ corresponds to pure derivative control.

Pure derivative control cannot be implemented. The range of ϕ_A is therefore $-\pi/2 \leq \phi_A \leq \varphi_0$ where φ_0 is about $\pi/3$ or 60°. With the Ziegler-Nichols frequency response method, it follows that:

$$\begin{aligned} G_c(j\omega_\pi) &= 0.6K_c\left(1 + j\left(\omega_\pi T_d - \frac{1}{\omega_\pi T_i}\right)\right), \\ &= 0.6K_c(0.6 + 0.28j). \end{aligned}$$

Thus, the Ziegler-Nichols frequency response method can be interpreted as a procedure which finds the control parameters so that the point where the Nyquist curve intersects the negative real axis is moved to $-0.6 - 0.28j$. This corresponds to a phase advance of $5\pi/36$ at ω_π.

2.4.3 The modified generalized frequency response method

This method is a general version of the Ziegler-Nichols frequency response method. Other points of the Nyquist curve can be selected to be moved to specified positions. In this way, it is possible to obtain design methods where the specifications are given in terms of amplitude or phase margins.

A general formulation is to start with a given point A of the Nyquist curve of the process:

$$G_p(j\omega) = k_A e^{j(-\pi+\phi_A)},$$

and to find a controller so that this point is moved to point B at $k_B e^{j(-\pi+\phi_B)}$. A pure amplitude margin design corresponds to $\varphi_s = 0$ and $r_s = 1/A_m$ where A_m is the amplitude margin; a pure phase margin design corresponds to $k_B = 1$ and $\phi_B = \phi_m$ where ϕ_m is the specified phase margin. The basic Ziegler-Nichols frequency response method is thus a specific case corresponding to $k_B = 0.66$ and $\phi_B = 0.44$.

The controller should thus be chosen so that:

$$k_C = \frac{k_B}{k_A},$$
$$\phi_C = \phi_B - \phi_A.$$

Simple calculations give

$$K_c = \frac{k_B \cos(\phi_B - \phi_A)}{k_A},$$
$$\omega T_d - \frac{1}{\omega T_i} = \tan(\phi_B - \phi_A).$$

The gain K_c is uniquely given. However, only one equation determines the parameters T_i and T_d. An additional condition must thus be introduced to determine these parameters uniquely. A common method is to specify a constant relation between T_i and T_d, i.e.,

$$T_d = \alpha T_i,$$

where a common choice is $\alpha = 0.25$. Straightforward calculations now give the parameters T_i and T_d as:

$$T_d = \frac{1}{2\omega}\left(-\tan(\phi_A - \phi_B) + \sqrt{4\alpha + \tan^2(\phi_A - \phi_B)}\right),$$

$$T_i = \frac{1}{\alpha}T_i$$

For systems where the amplitude and the phase of the transfer function decreases monotonously, the recommended choice of $k_B = 0.5$ and $\phi_B = \pi/4$ guarantees an amplitude margin of at least two and a phase margin of at least 45°.

The Ziegler-Nichols frequency response methods, basic or modified, are all based on the idea of moving one point on the Nyquist curve to a desired position. The terms phase margin and amplitude margin also define one point on the Nyquist curve. In most cases, these simple design rules are sufficient, but there are exceptional situations.

Fig. 2.4 shows the Nyquist curves of three systems having the same phase margin, $\phi_m = \pi/4$. This means that all the Nyquist curves pass through the point $s = -0.707 - 0.707j$. However, the step responses clearly demonstrate that the transient behavior of the control loop is also influenced by other points of the Nyquist curve.

More elaborate shaping (e.g., Tan et. al. 1996) of the compensated frequency response would resolve the problem mentioned above. Such methods involve knowledge of more than one points of the process frequency response and an iteration process to check the frequency response at other points, other than the one specified, to ensure good behavior in the compensated frequency response.

2.5 The Stability Limit Method

This method is very simple, and no instruments are necessary for data logging and analysis. However, it has its limitations. The operator tunes the controller by adjusting the PID parameters according to the following scheme:

1. Set the controller to automatic control, with the I and D parts disconnected.

2. Turn up the gain K_c until the control is unsteady. Then turn the gain down to half this value.

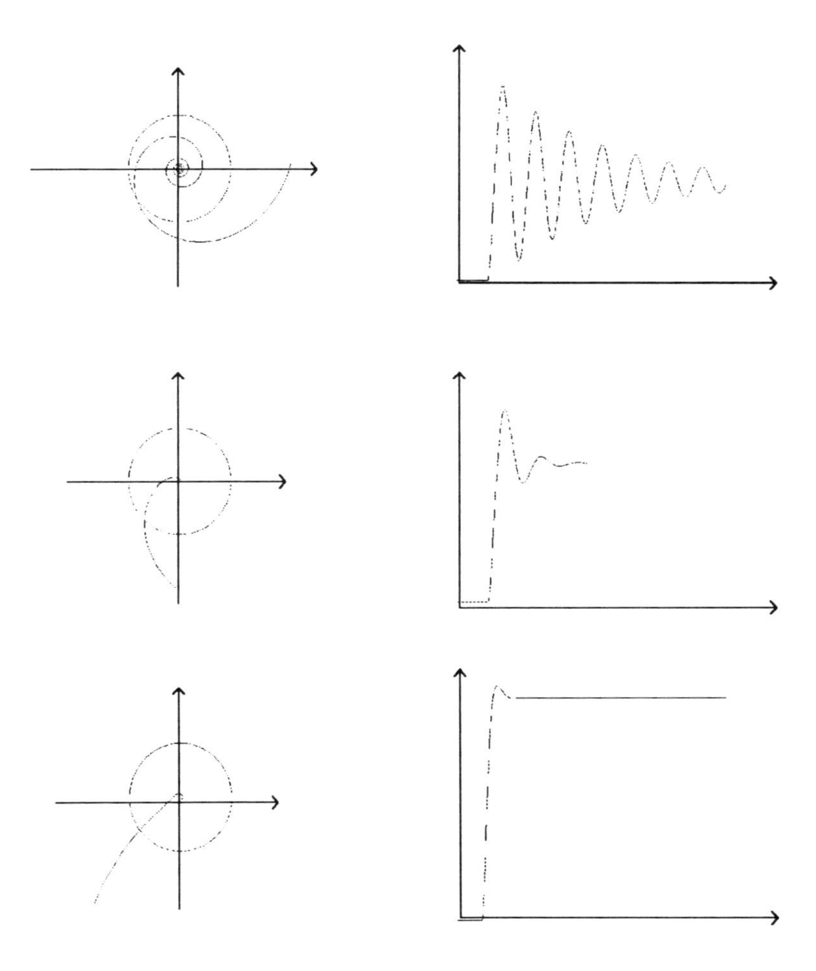

Fig. 2.4. Systems with similar phase margins but dissimilar step response

3. If the integral effect is required - turn down the integral time T_i until the control becomes unsteady. Then turn up the integral time to twice this value.

4. If the derivative effect is required - turn up the derivative time T_d until control becomes unsteady. Then turn the derivative time down to half.

The method is based on successively searching for the stability limits for the respective sections in the controller. The method is very coarse, however, and should be regarded as a way of getting rapidly into a more-or-less good range. The method also has the same disadvantage as the Ziegler-Nichols frequency method, namely that the stability limit for the control loop is engaged during the tuning procedures.

An important advantage is that the procedure is independent of the PID controller structure - parallel or series. The calibration of the scales for the PID parameters does not need to be precise either, because the settings are made without depending on any process model to determine the controller parameters.

A more systematic stability margin method is proposed by Luyben and Luyben (1997). The steps are described below:

1. With the controller in the manual mode, the I and D parts of the controller are disconnected, i.e., T_i is set at maximum and T_d at the minimum.
2. Set K_c at a low value, say 20%.
3. Put the controller in the automatic mode.
4. Make a small set point or load change and observe the response of the controlled variable. The gain is low, so the response will be sluggish.
5. Increase K_c by a factor of two and make another small change in set point or load.
6. Keep increasing K_c, repeating step 5 until the loop becomes very underdamped and oscillatory. The gain at which this occurs is called the ultimate gain.
7. Reduce K_c to half this ultimate value.
8. Now start bringing in integral action by reducing T_i by factors of 2, making small disturbances to see the effect.
9. Find the value of T_i that makes the loop very under-damped, and set T_i at twice this value.
10. Start bringing in derivative action by increasing T_d. Load changes should be used to disturb the system, and the derivative should act on the process variable signal. Find the value of T_d which gives the tightest control without amplifying the noise in the process variable signal.
11. Increase K_c again by steps of ten percent until the desired specification on damping coefficient or overshoot is satisfied.

2.6 The Cohen-Coon Method

The Cohen-Coon method is very similar to the Ziegler-Nichols step response method. It is based on the same principle of the Ziegler-Nichols step response method to achieve quarter amplitude damping. The experiment is carried out in exactly the same way as for the Ziegler-Nichols step response method, based on a step response experiment.

Using the same notation as in Section 2.4.1, the tuning rules are given in Table 2.4. The Cohen-Coon method was developed for the PID controller of the parallel type. However, the general expression for translating between parallel type to series type may be used to obtain the series type parameters.

Table 2.4. PID parameters according to the Cohen-Coon method

Controller	K	T_i	T_d
P	$\frac{1}{K_p}(0.35 + \frac{1}{\Theta})$	-	-
PI	$\frac{1}{K_p}(0.083 + \frac{0.9}{\Theta})$	$\frac{3.3+0.31\Theta}{1+2.2\Theta}L$	-
PD	$\frac{1}{K_p}(0.16 + \frac{1.24}{\Theta})$	-	$\frac{0.27-0.088\Theta}{1+0.13\Theta}L$
PID	$\frac{1}{K_p}(0.25 + \frac{1.35}{\Theta})$	$\frac{2.5+0.46\Theta}{1+0.61\Theta}L$	$\frac{0.37}{1+0.19\Theta}L$

The Cohen-Coon formulae are not as well-known as the Ziegler-Nichols rules. One reason could be that they are more demanding in calculation. Comparing Tables 2.4 and 2.2, it may be seen that the methods are quite similar for small values of Θ, i.e., when the process has a short dead-time in relation to the time constant. Ziegler-Nichols chose to connect the integral and derivative times completely to the dead-time in the process, while Cohen-Coon adjust the times according to the particular relationship between the dead-time and the time constant. This is the marked difference between the two methods.

In both methods, the controller gain is a function of this relationship. Because processes with large differences in the relationship between dead-time and time constant have marked differences in dynamic behavior, it could be expected that the Cohen-Coon method would work better than the Ziegler-Nichols method. For example, for very long dead-times (large value of Θ), the derivative time tends towards zero in the PID controller. This is correct, as the derivative part should not be used when the dead-time is long as it is more of a hindrance than help.

2.7 The Tyreus-Luyben Method

The Tyreus-Luyben method is quite similar to the Ziegler-Nichols frequency response method but it gives more conservative settings (higher closed-loop damping coefficient). The method sacrifices performance for robustness. The method also uses the ultimate gain k_π and ultimate frequency $\omega_\pi = \frac{2\pi}{T_\pi}$. Table 2.5 contains the parameter computation formulae.

Table 2.5. PID parameters according to Tyreus-Luyben method

Controller	K	T_i	T_d
P	-	-	-
PI	$\frac{K_\pi}{3.2}$	$2.2T_\pi$	-
PID	$\frac{K_\pi}{2.2}$	$2.2T_\pi$	$\frac{T_\pi}{6.3}$

CHAPTER 3
MODERN DESIGNS

3.1 Introduction

The requirements on a control system may include many factors such as response to command signals, insensitivity to measurement noise and process variations, and rejection of load disturbances. The design of a control system also involves aspects of process dynamics, actuator saturation, and disturbance characteristics. It may, therefore, seem surprising that a controller as simple as the PID regulator can work so well generally. The general empirical observation is that most industrial processes can be controlled reasonably well with PID control provided that the demands on the performance of the control are not too high. Even then, for certain difficult classes of systems to be elaborated in the next section, PID control may not yield satisfactory performance. For these systems, a more complex control structure may be necessary, or new ways of tuning the PID controllers beyond the classical methods should be explored. This chapter covers several modern approaches towards PID design method under more performance-critical situations.

3.2 Constraints of Classical PID Control

Control of systems with a dominant time-delay are notoriously difficult. It is also a topic on which there are many different opinions concerning the merit of PID control. There seems to be a general agreement that derivative action does not help much for processes with dominant time-delays. For open-loop stable processes, the response to command signals can be improved substantially by introducing a suitable dead-time compensation scheme such as the Smith predictor controller . The load disturbance rejection can also be improved to some degree because a dead-time compensator usually allow a larger loop gain than a PID controller. Systems with dominant time-delays are thus candidates for more sophisticated control or PID tuning.

Systems with oscillatory modes that occur when there are inertias and compliances is another case where PID control may not be sufficient. There are several approaches to control of these systems. In the so-called notch filter approach, no attempt is made to dampen the oscillatory modes, but an effort is made to reduce the signal transmission through the regulator by a filter that drastically reduces signal transmission at the resonant frequency. A PID controller may be used when there is only one dominant oscillatory mode. Notch filter action can be achieved by a judicious choice of the controller parameters. In this case, parameters T_i and T_d should, however, be chosen so that the numerator has complex roots.

Uncertain systems are yet another class of difficult processes for PID control. For systems with small parameter variations , simple detuning of the PID control may suffice. For systems with larger parameter variations, it is possible to design linear controllers that allow operation over a wide parameter range. Such controllers are, however, often of high-order.

The control of process variables that are closely related to important quality variables may be of a significant economic value. In such control loops, it is frequently necessary to select the controller with respect to the disturbance characteristics. This then often leads to strategies that are not of the PID type, since a PID controller has limited complexity and it cannot model disturbance behavior in general.

3.3 Pole Placement Design

3.3.1 PI control of first-order systems

Suppose the process can be described by the following first-order model:

$$G_p(s) = \frac{K_p}{1 + sT}, \tag{3.1}$$

which has only two parameters, the process gain K_p and the time constant T. By controlling this process with the PI controller,

$$G_c(s) = K_c \left(1 + \frac{1}{sT_i}\right),$$

a second-order closed-loop system is obtained:

$$G_{yr}(s) = \frac{G_p(s)G_c(s)}{1 + G_p(s)G_c(s)}.$$

The two closed-loop poles can be chosen arbitrarily by a suitable choice of the gain (K_c) and the integral time (T_i) of the controller. This is seen as follows. The poles are given by the characteristic equation, i.e., the equation

$$1 + G_p(s)G_c(s) = 0.$$

The characteristic equation becomes:

$$s^2 + s\left(\frac{1}{T} + \frac{K_pK_c}{T}\right) + \frac{K_pK_c}{TT_i} = 0.$$

Now suppose that the desired closed-loop poles are characterized by their relative damping ζ and their frequency ω. The desired characteristic equation then becomes

$$s^2 + 2\zeta\omega s + \omega^2 = 0.$$

Making the coefficients of these two characteristic equation equal gives two equations for determining K_c and T_i, from which the PI parameters are obtained:

$$K_c = \frac{2\zeta\omega T - 1}{K_p},$$
$$T_i = \frac{2\zeta\omega T - 1}{\omega^2 T}.$$

Notice that in order to have positive controller gains, it is necessary that the chosen bandwidth (ω) be larger than $\frac{1}{2\zeta T}$. Also notice that if ω is large, the integration time T_i is given by:

$$T_i = \frac{2\zeta}{\omega}.$$

It is thus independent of the process dynamics for large ω. There is no formal upper bound to the bandwidth. However, a simplified model like (3.1) will not hold for higher frequencies. The upper bound on the bandwidth is therefore determined by the validity of the model.

3.3.2 PID control of second-order systems

Suppose the process is characterized by the second-order model:

$$G_p(s) = \frac{K_p}{(1 + sT_1)(1 + sT_2)}. \tag{3.2}$$

This model has three parameters. By using a PID controller, which also has three parameters, it is possible to arbitrarily place the three poles of the closed-loop system. The transfer function of the PID controller can be written as

$$G_c(s) = \frac{K_c(1 + sT_i + s^2 T_i T_d)}{sT_i}.$$

The characteristic equation of the closed-loop system becomes:

$$s^3 + s^2\left(\frac{1}{T_1} + \frac{1}{T_2} + \frac{K_p K_c T_d}{T_1 T_2}\right) + s\left(\frac{1}{T_1 T_2} + \frac{K_p K_c}{T_1 T_2}\right) + \frac{K_p K_c}{T_i T_1 T_2} = 0.$$

A suitable closed-loop characteristic equation of a third-order system is

$$(s + \alpha\omega)(s^2 + 2\zeta\omega s + \omega^2) = 0,$$

which contains two dominant poles with relative damping (ζ) and frequency (ω), and a real pole located at $-\alpha\omega$. Identifying the coefficients in these two characteristic equations gives

$$K_c = \frac{T_1 T_2 \omega^2 (1 + 2\zeta\alpha) - 1}{K_p},$$

$$T_i = \frac{T_1 T_2 \omega^2 (1 + 2\zeta\alpha) - 1}{T_1 T_2 \alpha \omega^3},$$

$$T_d = \frac{T_1 T_2 \omega(\alpha + 2\zeta) - T_1 - T_2}{\omega^2 T_1 T_2 (1 + 2\zeta\alpha) - 1}.$$

Note that pure PI control is obtained for

$$\omega = \omega_c = \frac{T_1 + T_2}{(\alpha + 2\zeta) T_1 T_2}.$$

Notice that the choice of ω may be critical. The derivation time is negative for $\omega < \omega_c$. The frequency (ω_c) thus gives a lower bound to the bandwidth, Also notice that the gain increases rapidly with ω. The upper bound to the bandwidth is given by the validity of the simplified model (3.2).

3.3.3 General case

Pole placement may be extended to general linear systems. However, depending on the order of the system, more closed-loop poles need to be specified. Various pole locations may be chosen. In the Butterworth configuration, the roots of the characteristic polynomials are placed symmetrically in a circle. In the Bessel configurations, filters are designed to preserve the shape of the wave form. The order of the controllers will correspondingly increases with the complexity of the model. To still use PID controllers, these high-order models have to be approximated by a reduced-order model in the form of a first or second-order transfer function. Several ways to perform these approximations are given in Chapter 4.

3.4 Dominant Pole Placement

Another way to resolve the problem related to the need for multiple closed-loop poles is to use the concept of dominant poles (Astrom and Hagglund, 1995). The idea is to assign only a few poles which will be sufficiently representative of the actual process. With this method, it is thus possible to use simple controllers for complex processes. The Cohen-Coon method is in fact based on positioning dominant poles to achieve a quarter amplitude decay ratio. The dominant pole placement method is illustrated below for PI and PID controllers, but the principle is easily extended to other PID permutations.

PI Control. PI control has two control parameters. Consequently, only two dominant poles need to be assigned irregardless of the process. Consider a process with the transfer function $G_p(s)$. The PI controller is parameterized as

$$G_c(s) = K_c + \frac{K_c}{T_i s}.$$

The closed-loop characteristic equation is:

$$1 + \left(K_c + \frac{K_c}{T_i s}\right) G_p(s) = 0. \tag{3.3}$$

The closed-loop poles are specified as a conjugate pair at

$$p_{1,2} = \omega_n e^{j(\pi \pm \phi)},$$

where $\cos\phi$ is the equivalent damping ratio and ω_n is the natural frequency. Therefore, the desired closed-loop characteristic equation is

$$1 + \left(K_c + \frac{K_c}{T_i p_1}\right) G_p(p_1) = 0. \tag{3.4}$$

Define

$$G_p\left(\omega_n e^{j(\pi-\phi)}\right) = M(\omega_n) e^{j\bar{\phi}(\omega_n)}.$$

Note that $G_p\left(\omega_n e^{j(\pi-\phi)}\right)$ represents the values of the transfer function on the ray $e^{j(\pi-\phi)}$. This value may be obtained via a modified relay experiment to be described in Chapter 4.

Comparing (3.3) and (3.4), the control gains may be computed as:

$$K_c = -\frac{\sin(\bar{\phi}(\omega_n) + \phi)}{M(\omega_n)\sin\phi},$$

$$K_i = -\frac{K_c M(\omega_n)\sin\phi}{\omega_n \sin(\bar{\phi}(\omega_n))}.$$

For positive values of both parameters, the natural frequency must be selected such that

$$\phi < -\bar{\phi}(\omega_n) < \pi$$

PID Control. Consider the PID controller parameterized as:

$$G_c(s) = K_c + \frac{K_c}{T_i s} + K_c T_d s.$$

With three parameters, it is necessary to specify three poles of the closed-loop system. A real pole and a conjugate pair is chosen at the following locations:

$$\begin{aligned} p_{1,2} &= \omega_n e^{j(\pi \pm \phi)}, \\ p_3 &= -\alpha\omega_n. \end{aligned}$$

Introduce the following definitions:

$$G_p\left(\omega_n e^{j(\pi-\phi)}\right) = M(\omega_n) e^{j\bar{\phi}(\omega_n)},$$

and

$$G_p(-\alpha\omega_n) = -M_\alpha(\omega_n).$$

Substituting $s = p_1$, $s = p_2$ and $s = p_3$ into the characteristic equation

$$1 + G_p(s)G_c(s) = 0,$$

give the control gains as:

$$\begin{aligned} K_c &= -\frac{\alpha^2 M_\alpha(\omega_n)\sin(\phi+\bar{\phi}) + M_\alpha(\omega_n) sin(\phi-\bar{\phi}) + \alpha M(\omega_n)\sin 2\phi}{M(\omega_n)M_\alpha(\omega_n)(\alpha^2 - 2\alpha\cos\phi + 1)\sin\phi}, \\ T_i &= \frac{K_c M(\omega_n)M_\alpha(\omega_n)(\alpha^2 - 2\alpha\cos\phi + 1)\sin\phi}{(-\alpha\omega_n)(\alpha\sin\phi + M_\alpha(\omega_n) sin(\phi-\bar{\phi}) + \alpha\sin\bar{\phi})}, \\ T_d &= -\frac{\alpha M(\omega_n)\sin\phi + M_\alpha(\omega_n)(\alpha sin(\phi+\bar{\phi}) - \sin\bar{\phi})}{K_c\omega_n M(\omega_n)M_\alpha(\omega_n)(\alpha^2 - 2\alpha\cos\phi + 1)\sin\phi}. \end{aligned}$$

3.5 Gain and Phase Margin Design I: PI Controller

Gain and phase margins are classical control loop specifications associated with the frequency response technique (Franklin and Powell, 1986). They reflect on the performance and stability of the system and are widely used for controller designs. There are a number of PID tuning formulae expressed in terms of the gain and phase margins. However, they either cannot achieve exactly both gain and phase margin specifications (Astrom and Hagglund, 1995), or make simplifications on the structure of the process such as the assumption of the plant being a first-order plus dead-time one (Ho et al., 1995). For processes of higher orders, users can only specify the gain and phase margin as if the processes are of the first-order plus dead-time type. However, the same gain and phase margin may generate very different performance for different kinds of process dynamics. This problem has not been addressed in the literature and no guideline can be found on how to set the gain and phase margins to have the desired closed-loop performance for a general linear process.

The major difficulties in controller designs that meet exact gain and phase margins lie in the non-linearity and solvability of the problem. In this section, a simple and straightforward controller design based on a graphical approach is presented which can simultaneously achieve exact gain and phase margin for a general linear plant, without making any structural assumption. The core of the method is about plotting two graphs similar to Nyquist curves and finding the intersection between them. Each of the intersections then generates a solution for the problem. With the method, not only is the existence of solution apparent, but all solutions, if they exist, can be clearly read off the graphs. In cases where there is no solution, a compromised solution can still be found to meet either the exact gain margin or the exact phase margin specification. Situations when the accomplishment of the specified gain and phase margins do not provide satisfactory response are accounted for gracefully by modifying the specifications for improvements.

3.5.1 The design method

Suppose that the transfer function $G_p(s)$, or the frequency response $G_p(j\omega)$, of a linear process is available and the single loop controller configuration as shown in Fig. 3.1 is adopted. A PI controller with the transfer function:

$$G_c(s) = K_p + \frac{K_i}{s}, \tag{3.5}$$

is employed to control the process. Assume that the control system specifications are given in terms of gain margin A_m and phase margin ϕ_m. These imply that

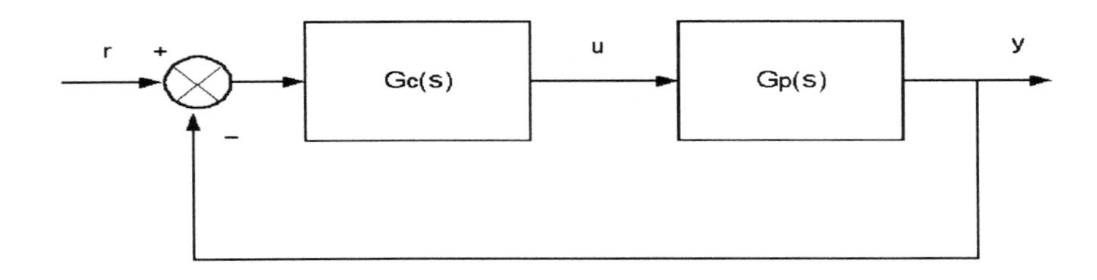

Fig. 3.1. Single-loop feedback control system

$$G_p(j\omega_p)\left(K_p - j\frac{K_i}{\omega_p}\right) = -\frac{1}{A_m}, \tag{3.6}$$

$$G_p(j\omega_g)\left(K_p - j\frac{K_i}{\omega_g}\right) = -e^{j\phi_m}, \tag{3.7}$$

where ω_p and ω_g are the phase and gain crossover frequencies of the loop respectively. Now the tuning objective is to determine the controller parameters K_p and K_i such that the given gain and phase margins (A_m and ϕ_m) are achieved, i.e., (3.6) and (3.7) are satisfied.

It is noted that there are a total of four unknowns, namely, K_p, K_i, ω_p and ω_g in (3.6) and (3.7). Since both equations are complex, they can be further broken down to four real equations. Since the number of unknowns equals the number of real equations, the number of solutions will be finite. Intuitively, it may be expected that since the equations for the controller parameters K_p and K_i have been readily constructed and they are known to have limited number of solutions, the problem can be easily solved. Unfortunately this is not the case. The main difficulties lie in the non-linearities of the equations which are further complicated by the coupling among them. Usually, numerical methods have to be used and an iterative process is needed to compute the solution. Even then, the solution is not straightforward. Convergence, initial estimate and long search time problems arise and worst of all, the existence of a solution is not certain. In the case when there is no solution, the iteration procedure will diverge or it will be trapped in an infinite loop.

For non-negative parameters, the controller (3.5) can only contribute a phase angle in the range from 0 to $-\frac{\pi}{2}$. Following this constraint, (3.6) and (3.7) lead to

$$\frac{-\pi}{2} < \angle G_p(j\omega_p) < -\pi, \tag{3.8}$$

$$\frac{-\pi}{2} < \angle G_p(j\omega_g) < -\pi + \phi_m. \tag{3.9}$$

Splitting (3.6) and (3.7) into their respective real and imaginary parts yields:

$$K_p = Re\left(\frac{-1}{A_m G_p(j\omega_p)}\right) = Re\left(\frac{-e^{j\phi_m}}{G_p(j\omega_g)}\right), \tag{3.10}$$

$$K_i = -\omega_p Im\left(\frac{-1}{A_m G_p(j\omega_p)}\right) = -\omega_g Im\left(\frac{-e^{j\phi_m}}{G_p(j\omega_g)}\right), \tag{3.11}$$

Since the frequency points ω_p and ω_g are unknown, the following two complex functions are defined:

$$f_p(\omega) = Re\left(\frac{-1}{A_m G_p(j\omega_p)}\right) \quad -j\omega Im\left(\tfrac{-1}{A_m G_p(j\omega)}\right); \quad -\tfrac{\pi}{2} < \angle G_p(j\omega) < -\pi, \tag{3.12}$$

$$f_g(\omega) = Re\left[\frac{-e^{j\phi_m}}{G_p(j\omega)}\right] \quad -j\omega Im\left[\tfrac{-e^{j\phi_m}}{G_p(j\omega)}\right]; \quad -\tfrac{\pi}{2} < \angle G_p(j\omega) < -\pi + \phi_m. \tag{3.13}$$

The graphs are plotted in the same complex plane. An intersection of these two graphs will indicate that they have the same real and imaginary parts, and thus both (3.10) and (3.11) will be satisfied so that the intersection is a solution to (3.6) and (3.7). Therefore, the two graphs will intersect at as many points as the number of solutions to (3.6) and (3.7) that provide a set of values for K_p, K_i, and the corresponding ω_p and ω_g. No solution exists if there is no intersection point.

It is the general observation from simulation study that when the intersection points are to the right of the peak of $f_g(\omega)$, a small integral gain K_i (relative to the corresponding proportional gain K_p) will result. In that case, due to insufficient integral effort, a slow response will result. It follows from Astrom and Hagglund (1995) that if the asymptotic behavior of the loop transfer function at low frequencies is close to or smaller than and appearing to the left of the -0.5 point, the integral action is sufficient. Hence, to guarantee an adequate K_i for reasonable closed-loop response, only the intersection point leftward to the peak of $f_g(\omega)$ will be chosen as a solution, since such a solution meets the above rule and also generates satisfactory response (Example 3.1 in the next section). This is accomplished by starting the search for the intersection point from the peak of $f_g(\omega)$ and going leftwards.

If the two graphs do not intersect before the peak of $f_g(\omega)$, an approximate solution has to be sought. It is possible to achieve either exact gain margin or exact phase margin with some change to the remaining margin. This can be accomplished by working on the same graphs without re-specification or re-plotting. It is easily seen from (3.12) that increasing the gain margin specification will shrink $f_p(\omega)$. Likewise, increasing the phase margin will cause a compression in $f_g(\omega)$. Since the two graphs $f_p(\omega)$ and $f_g(\omega)$ do not meet in the range, one must be lower than the other. If a solution point is instead picked from the lower graph, intersection can be forced by shrinking the other graph. This is equivalent to increasing the value of the margin corresponding to the upper graph. The system will then exactly meet one margin and achieve an extra degree of conservativeness for the other margin. Conversely, if the solution point is chosen from the upper graph, the system will meet the margin corresponding to it, but attain less than the specified requirement for the other by expanding the lower graph. In any case, it is advisable to achieve extra margin from the stability point of view at the expense of some loss in speed. Further, the change to one margin should be minimum in order to achieve the given specifications as closely as possible. Thus, the design will shrink the upper graph by the minimum amount so as to create a forced intersection, and the intersection will yield the solution. In addition, the solution should lies to the left of the peak of $f_g(\omega)$ for a good response.

To summarize the above discussions, the following rules are developed for the determination of the solution:

RULE 1

- *Start the search for the intersection point from the peak point on $f_g(\omega)$ leftwards. If there is an intersection point, it will be taken as the solution point.*

RULE 2

- *If no intersection is obtained this way from the search, shrink the upper graph by the minimum amount so as to generate a forced intersection which is the peak point on $f_g(\omega)$, or to the left of the peak. This forced intersection will be taken as the solution point.*

In nearly all cases, if $f_g(\omega)$ lies above $f_p(\omega)$, the process is oscillatory, and the intersection point, obtained by shrinking $f_g(\omega)$,will be the point on $f_p(\omega)$ with $K_p = 0$. On the other hand, if $f_g(\omega)$ lies below $f_p(\omega)$, the process is non-oscillatory and the intersection point, obtained by shrinking $f_p(\omega)$, will be the peak point of $f_g(\omega)$.

Accordingly, the tuning procedure is as follows:

- If the graphs $f_g(\omega)$ and $f_p(\omega)$ intersect at a point to the left of $f_g(\omega)$, take it as the solution point.
- Otherwise, if $f_g(\omega)$ lies above $f_p(\omega)$, choose the point on $f_p(\omega)$ with $K_p = 0$ as the solution point. If $f_g(\omega)$ lies below $f_p(\omega)$, then choose the peak point as the solution point.
- Set the PI parameters K_p and K_i as the real and imaginary parts of the solution point respectively.

3.5.2 Simulation study

Some examples will be provided in this section to demonstrate the use of the method (thenceforth referred to as Exact-GPM). Comparisons will be made with the method of Ho and co-workers (GPM), (Ho et al., 1995) where the process is modeled as a first-order plus dead-time process. The method is designed to meet user-defined gain and phase margins, which is similar to the objective of the Exact-GPM method, thus setting a fair base for comparison. The gain and phase margins will be chosen to be three and 60 degrees respectively in the following examples for both methods. Step responses obtained from the Exact-GPM method are plotted in solid lines while those obtained from Ho's method are shown in dashed lines.

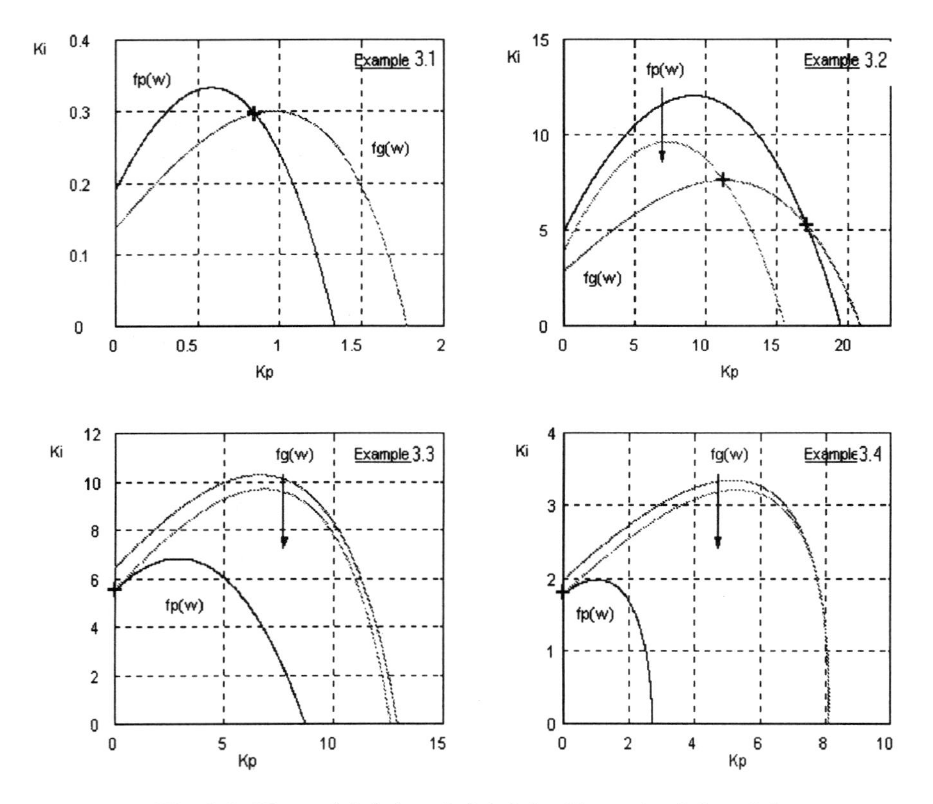

Fig. 3.2. Plots of $f_p(\omega)$ and $f_g(\omega)$ for Examples 3.1 to 3.4

Example 3.1:

Consider a non-oscillatory high-order process:

$$G_p(s) = \frac{1}{(s+1)^4}.$$

The graphs of $f_p(\omega)$ and $f_g(\omega)$ are shown in Fig. 3.2. The search process starts leftwards from the peak of $f_g(\omega)$ and the intersection point is found to be at 0.848+j0.297. Therefore, the resulting PI controller is obtained as:

$$G_c(s) = 0.848 + \frac{0.297}{s}.$$

Gain and phase margins of $A_m = 3.0$ and $\phi_m = 59.9$ degrees are achieved. GPM method yields

$$G_c(s) = 1.112 + \frac{0.287}{s}.$$

The gain and phase margins achieved in this case are $A_m = 2.6$ and $\phi_m = 60.9$ degrees respectively. Since the process can be well represented by the first-order plus dead-time model, the gain and phase margins and PI parameters achieved by GPM method are quite close to the Exact-GPM method. Similar control performance is hence expected. A comparison of the performance is shown in Fig. 3.3.

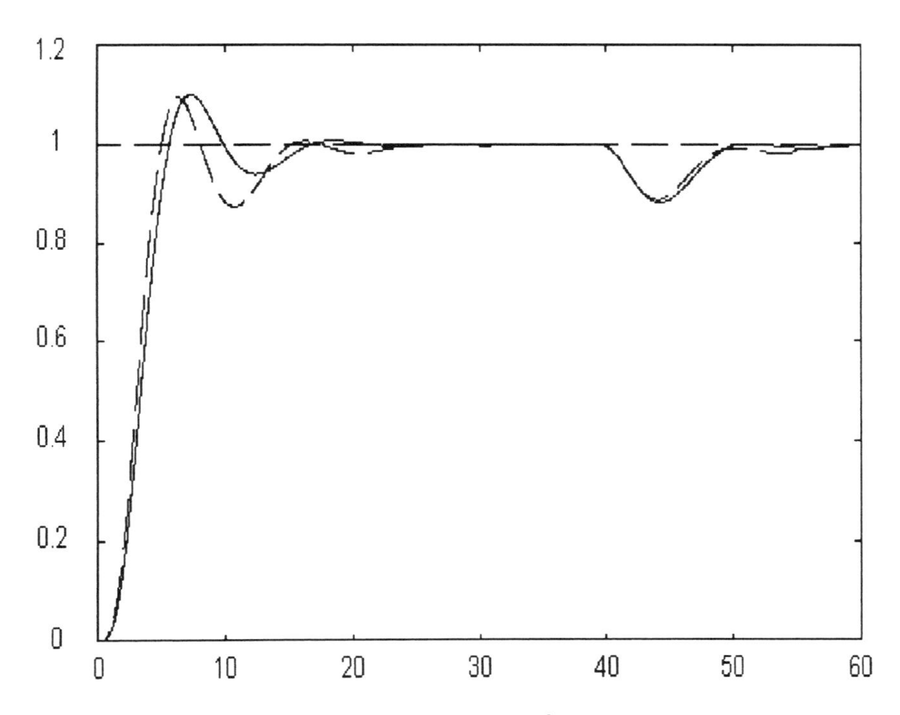

Fig. 3.3. Control performance with $G_p(s) = \frac{1}{(s+1)^4}$ (solid:- Exact-GPM, dashed:- GPM): Example 3.1

Example 3.2:

Consider another non-oscillatory process:

$$G_p(s) = \frac{1}{(s+1)(s+3)^2}e^{-0.1s}.$$

The graphs of $f_p(\omega)$ and $f_g(\omega)$ and are plotted in Fig. 3.4. An intersection point is found at 17.072+j5.223, but it is located beyond the peak of $f_g(\omega)$ and hence gives rise to a small K_i to K_p ratio. The corresponding Nyquist plot of the loop is shown in dotted line in Fig. 3.4. The asymptotic value of the loop gain at low frequency has a real part in the far right of the complex plane and thus fails to satisfy the rule of Astrom and Hagglund (1995) which suggests that the real part should be close to, or less than (on the left of) -0.5 for an adequate integral effort.

Although exact gain and phase margins can be achieved with this solution, a sluggish yet oscillatory response is obtained as shown in dotted line in Fig. 3.5. The solution is discarded. If the search is initiated from the peak of $f_g(\omega)$ leftwards, no intersection will be found. Hence to force an intersection, $f_p(\omega)$ is shrunk by increasing the gain margin to $A_m = 3.76$, and the two graphs now meet at 11.224+j7.607, which corresponds to the peak of $f_g(\omega)$. The PI controller using the Exact-GPM method is thus found to be:

$$G_c(s) = 11.224 + \frac{7.607}{s}.$$

In Fig. 3.4, the Nyquist plot for the loop with the modified controller is shown which is shifted left to give a negative real part at low frequency, thus achieving better integral control action. Design with GPM method yields:

$$G_c(s) = 17.527 + \frac{8.308}{s}.$$

The gain and phase margins are $A_m = 2.7$ and $\phi_m = 49.8$ degrees respectively. The closed-loop performances are together shown in Fig. 3.5, and it is observed that the Exact-GPM method gives a much better response in terms of a smaller overshoot and a shorter settling time.

Example 3.3:

Consider an oscillatory process given by:

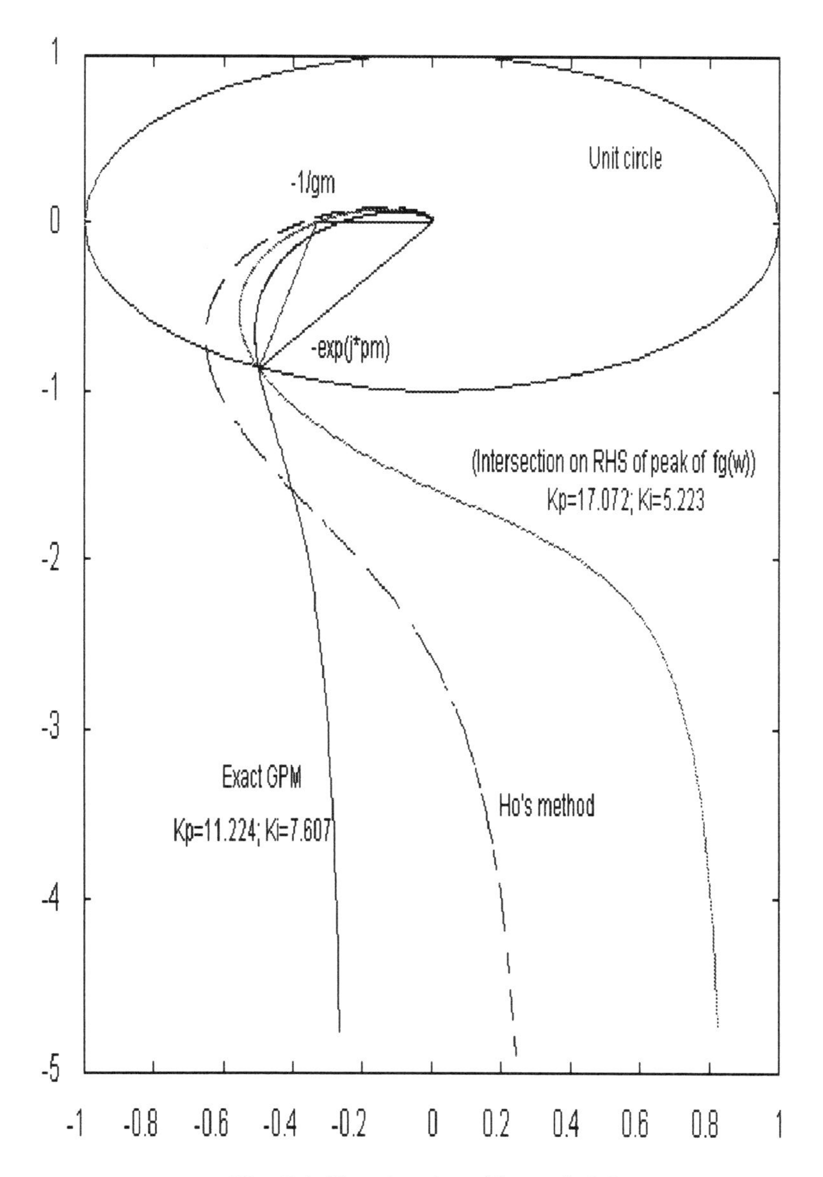

Fig. 3.4. Nyquist plots: Example 3.2

$$G_p(s) = \frac{1}{(s^2 + 2s + 4)(s + 3)} e^{-0.1s}.$$

Fig. 3.2 shows the graphs of $f_p(\omega)$ and $f_g(\omega)$. No intersection point can be found. Since $f_p(\omega)$ lies below $f_g(\omega)$, $f_g(\omega)$ is shrunk by raising the phase

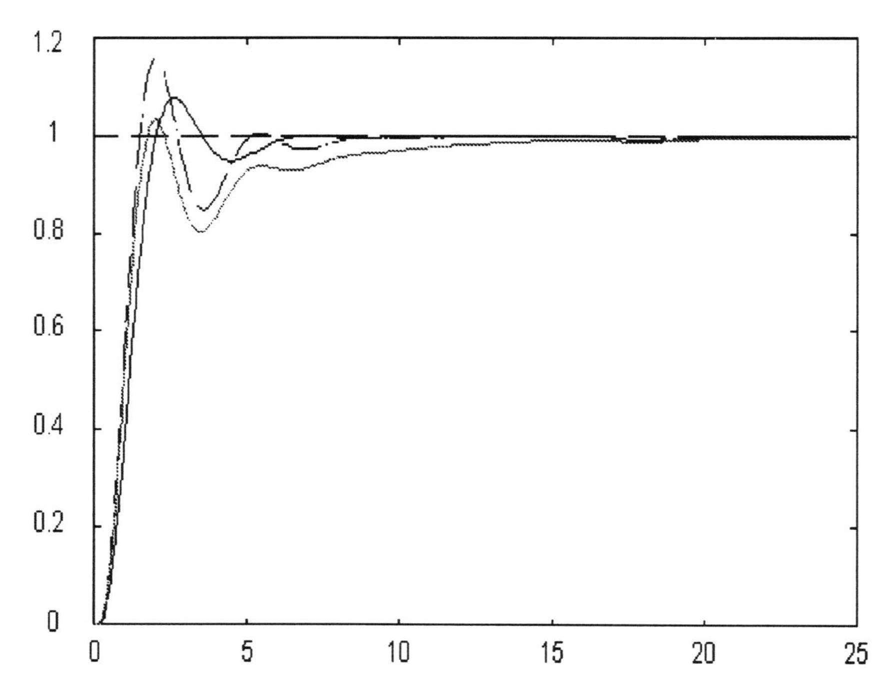

Fig. 3.5. Control performance with $G_p(s) = \frac{1}{(s+1)(s+3)^2}e^{-0.1s}$ (solid:- Exact-GPM, dashed:- GPM): Example 3.2

margin to $\phi_m = 64.6$ degrees to produce an intersection. The controller is determined to be:

$$G_c(s) = \frac{5.532}{s},$$

with $K_p = 0$. Design using GPM method gives:

$$G_c(s) = 5.920 + \frac{8.357}{s},$$

leading to gain and phase margins of $A_m = 2.3$ and $\phi_m = 73.1$ degrees respectively. The closed-loop step responses are shown in Fig. 3.6. Insignificant oscillation in the response is observed with the Exact-GPM method and significant improvement in performance is achieved.

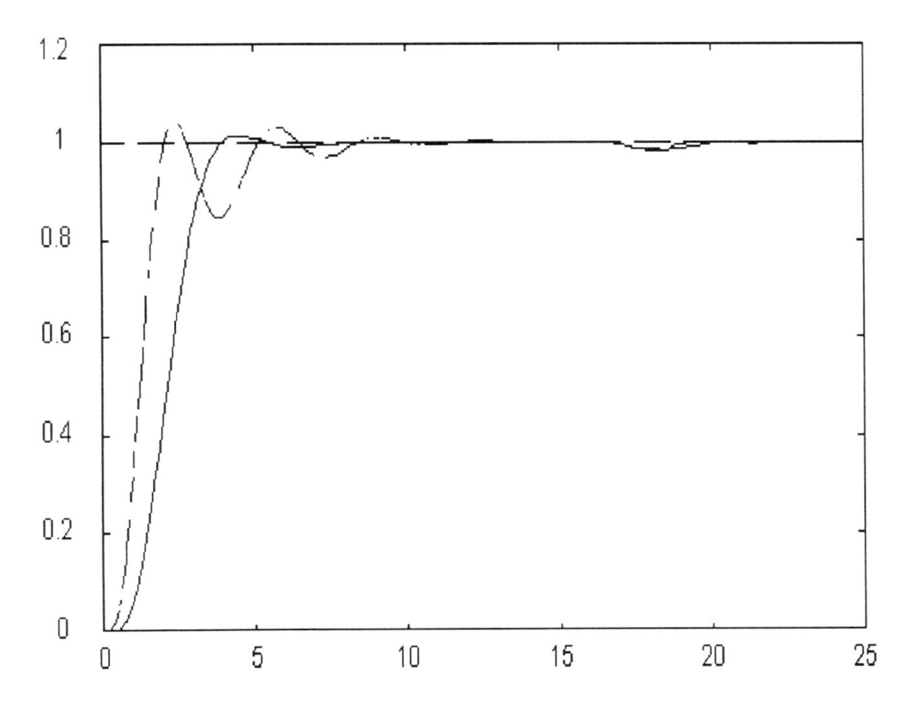

Fig. 3.6. Control performance with $G_p(s) = \frac{1}{(s^2+2s+4)(s+3)}e^{-0.1s}$ (solid:- Exact-GPM, dashed:- GPM): Example 3.3

Example 3.4:

Consider a high-order oscillatory process:

$$G_p(s) = \frac{1}{(s^2+2s+3)^3}e^{-s}.$$

Following the Exact-GPM method, it follows that

$$G_c(s) = \frac{1.796}{s},$$

with gain and phase margins achieved of $A_m = 3.0$ and $\phi_m = 63.0$ degrees respectively. GPM method provides no solution in this case. The step response is shown in Fig. 3.7, and satisfactory performance is observed.

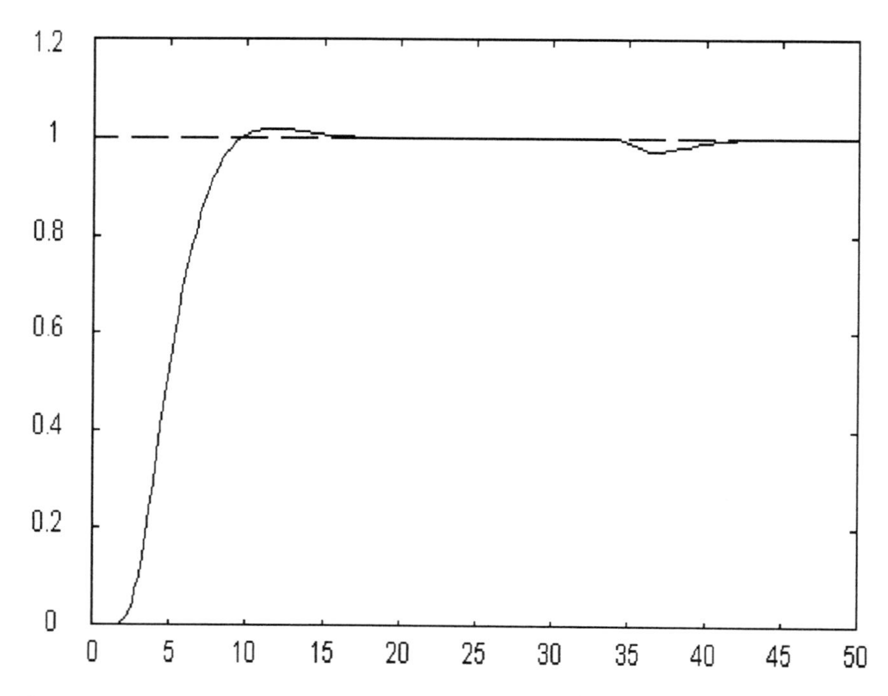

Fig. 3.7. Control performance with $G_p(s) = \frac{1}{(s^2+2s+3)^3}e^{-s}$ (solid:- Exact-GPM, dashed:- GPM): Example 3.4

3.6 Gain and Phase Margin Design II: PID Controller

Suppose the transfer function or the frequency response of a linear stable process is available and the single loop controller configuration as shown in Fig. 3.1 is adopted. A PID controller with the transfer function:

$$G_c(s) = K_p + \frac{K_i}{s} + K_d s, \tag{3.14}$$

is employed to control the process. Assume that the control system specifications are given in terms of gain margin A_m and phase margin ϕ_m, subject to $A_m > 1$ and $0 < \phi_m < \frac{\pi}{2}$. It follows from the margin definition:

$$G_p(j\omega_p)\left[K_p + j\left(K_d\omega_p - \frac{K_i}{\omega_p}\right)\right] = -\frac{1}{A_m}, \tag{3.15}$$

$$G_p(j\omega_g)\left[K_p + j\left(K_d\omega_g - \frac{K_i}{\omega_g}\right)\right] = -e^{j\phi_m}, \tag{3.16}$$

where ω_p and ω_g are the phase and gain crossover frequencies of the loop respectively. Now the tuning objective is to determine the controller parameters K_p, K_i and K_d so that the given gain and phase margins (A_m and ϕ_m) are achieved, i.e., (3.15) and (3.16) are satisfied.

It is noted that there are altogether five unknowns, namely K_p, K_i, K_d, ω_p and ω_g, in (3.15) and (3.16). Since the equations are complex, they may be broken down into four real equations. Since the number of unknowns exceeds the number of equations, there is an infinite number of solutions to (3.15) and (3.16), unless one extra constraint is added.

The additional equation can be introduced by considering the bandwidth of the system. The bandwidth of a process is defined as the frequency at which the process gain is reduced to 3dB below the gain at zero frequency. It is usually approximated by the phase crossover frequency, since frequencies below the phase crossover frequency constitute the most significant frequency range in controller design. In controller tuning, the closed-loop bandwidth should be carefully chosen. If it is too large , the control signal will saturate. If it is too small, sluggish response will result. It is commonly accepted in engineering practice that the closed-loop bandwidth should be close to the open-loop bandwidth. In the Exact-GPM method, it is chosen as:

$$\omega_p = \alpha\omega_c, \quad \alpha \in [0.5, 2], \tag{3.17}$$

where ω_c and ω_p are the phase crossover frequencies of the process G_p and G_pG_c respectively. The default value for α is chosen as $\alpha = 1$. ω_c is directly available from the process frequency response and it is the point that satisfying:

$$\angle G_p(j\omega_c) = -\pi. \tag{3.18}$$

It can be easily read off from plots of $G_p(j\omega)$ or found from data of $G_p(j\omega)$, since it is the lowest non-zero frequency point at which the imaginary part of equals to zero. With (3.17), ω_p is readily determined and the number of unknowns is reduced from five to four. (3.16) and (3.17) can then be solved to give

$$K_p = Re\left[\frac{-1}{A_m G_p(j\omega_p)}\right], \tag{3.19}$$

$$K_i = (X_p\omega_g - X_g\omega_p)\left(\frac{\omega_p}{\omega_g} - \frac{\omega_g}{\omega_p}\right)^{-1}, \tag{3.20}$$

$$K_d = \left(\frac{X_p}{\omega_g} - \frac{X_g}{\omega_p}\right)\left(\frac{\omega_p}{\omega_g}\frac{\omega_g}{\omega_p}\right)^{-1}, \tag{3.21}$$

where

$$X_p = Im\left[\frac{-1}{A_m G_p(j\omega_p)}\right], \tag{3.22}$$

$$X_g = Im\left[\frac{-e^{j\phi_m}}{G_p(j\omega_g)}\right], \tag{3.23}$$

and ω_g satisfies

$$Re\left[\frac{-e^{(j\phi_m)}}{G_p(j\omega_g)}\right] = Re\left[\frac{-1}{A_m G_p(j\omega_p)}\right], \tag{3.24}$$

which is equal to K_p.

In view of the above development, the design problem can be solved if and only if (3.24) admits a solution for ω_g. One notes that (3.24) can be equivalently put into:

$$\frac{|G_p(j\omega_p)|cos(\phi_m - \angle G_p(j\omega_g))}{|G_p(j\omega_g)|cos(\angle G_p(j\omega_p))} = \frac{1}{A_m}.$$

The point ω_g may be identified by searching downwards from the frequency $\omega = \omega_p$ towards $\omega = 0$ until (3.24) holds. Let

$$f(\omega) = \frac{|G_p(j\omega_p)|cos(\phi_m - \angle G_p(j\omega_g))}{|G_p(j\omega)|cos(\angle G_p(j\omega_p))}.$$

Then, (3.24) will have a solution for $\omega_g \in [0, \omega_p]$ if $f(0) < \frac{1}{A_m}$ and $f(\omega_p) > \frac{1}{A_m}$. As a PID controller provides a phase leag or lag of no more than $\frac{\pi}{2}$, (3.15) implies

$$-\frac{3\pi}{2} < \angle G_p(j\omega_p) < -\frac{\pi}{2}, \tag{3.25}$$

and $cos(\angle G_p(j\omega_p)) < 0$. For $0 < \phi_m < \frac{\pi}{2}$ and $A_m > 1$, it follows that $cos\phi_m > 0$ and,

$$f(0) = \frac{|G_p(j\omega_p)|cos(\phi_m)}{|G_p(j0)|cos(\angle G_p(j\omega_p))} < 0 < \frac{1}{A_m}.$$

On the other hand, the condition that $f(\omega_p) > \frac{1}{A_m}$ requires:

$$\frac{cos(\phi_m - \angle G_p(j\omega_p))}{cos(\angle G_p(j\omega_p))} > \frac{1}{A_m},$$

or

$$tan(\angle G_p(j\omega_p)) > \frac{1 - A_m cos\phi_m}{A_m sin\phi_m}.$$

Combining it with (3.25) yields

$$-\pi + arctan\left(\frac{1 - A_m cos\phi_m}{A_m sin\phi_m}\right) < \angle G_p(j\omega_p) < -\frac{\pi}{2}. \tag{3.26}$$

(3.26) is a condition which guarantees (3.24) to have a solution, and it can be easily verified with the given A_m ϕ_m and ω_p. For the default case of

$\alpha = 1$, where $\angle G_p(j\omega_p) = -\pi$, (3.26) reduces to $arctan\left(\frac{1-A_m cos\phi_m}{A_m sin\phi_m}\right) < 0$, or $A_m cos\phi_m > 1$, which is a very simple and direct test. Furthermore, for the most typical specifications where the gain and phase margins are set to $A_m = 3$ and $\phi_m = \frac{\pi}{3}$ (or 60 degrees) respectively, $A_m cos\phi_m = 1.5 > 1$ and there will always be a solution to (3.24).

To demonstrate how to use (3.26), consider a second-order plus dead-time process:

$$G_p(s) = \frac{1}{(s+1)(s+3)}e^{-s}.$$

The phase crossover frequency of the process is $\omega_c = 1.626$ rad/s. If the gain and phase margins are first set at $A_m = 1.5$ and $\phi_m = 1.396$ (or 80 degrees) respectively, (3.26) becomes $-2.677 < \angle G_p(j\omega_p) < -\frac{\pi}{2}$, and the corresponding ω_p satisfies $0.720 < \omega_p < 1.322$ rad/s. The choice of $\omega_p = \omega_c = 1.626$ rad/s will violate (3.26). One possibility to meet (3.26) is to reduce ω_p to $\omega_p = 1.3 rad/s$, then the solution to (3.24) is obtained as $\omega_g = 0.565$ rad/s. Another possibility, with $\omega_p = \omega_c$ unchanged, is to modify the gain and phase margins to $A_m = 3.0$ and $\phi_m = \frac{\pi}{3}$ (or 60 degrees), respectively. (3.26) then gives $-3.332 < \angle G_p(j\omega_p) < -\frac{\pi}{2}$, or $0.720 < \omega_p < 1.754$ rad/s. $\omega_p = \omega_c = 1.626$ rad/s lies in this range. The solution to (3.24) is then obtained as $\omega_g = 0.535$ rad/s.

Tuning Procedure

Given the plant $G_p(s)$ or $G_p(j\omega)$, the PID parameters can be tuned to meet both gain margin A_m and phase margin ϕ_m in the following way:

- Obtain the process phase crossover frequency ω_c from $G_p(j\omega)$.
- Determine the ratio of closed-loop to open-loop bandwidth, $\alpha(\alpha \in [0.5, 2]$with a default of 1) and set $\omega_p = \alpha\omega_c$.
- Check whether or not (3.26) is satisfied. If not, either reduce ω_p or modify the gain/phase margin to meet (3.26).
- Calculate K_p from (3.19).
- Search from $\omega = \omega_p$ down towards $\omega = 0$ for the frequency ω_g that satisfies (3.24).
- Compute K_i and K_d from (3.20) and (3.21).

If the frequency response of the process is not available, an estimate can be obtained by performing a step test or relay test on the process. The input and output data u(t), y(t) are collected from $t = 0$ until the steady state is reached at $t = t_s$ when both $y(t) = y_s(t)$ and $u(t) = u_s(t)$ for $t \geq t_s$ become periodic. The process frequency response is then calculated (Wang et al., 1999) using the formula:

$$G_p(j\omega) = \frac{Y_s(j\omega) + \triangle Y(j\omega)}{U_s(j\omega) + \triangle U(j\omega)}, \tag{3.27}$$

where

$$Y_s(j\omega) = \begin{cases} \frac{y_s}{j\omega}, step; \\ \frac{1}{1-e^{-j\omega T_c}} \int_0^{T_c} y_s(t)e^{-j\omega t}dt, relay, \end{cases} \tag{3.28}$$

and

$$U_s(j\omega) = \begin{cases} \frac{u_s}{j\omega}, step; \\ \frac{1}{1-e^{-j\omega T_c}} \int_0^{T_c} u_s(t)e^{-j\omega t}dt, relay, \end{cases} \tag{3.29}$$

and $\triangle y(t) = y(t) - y_s(t)$,$\triangle u(t) = u(t) - u_s(t)$. The values of $\triangle Y(j\omega)$ and $\triangle U(j\omega)$ are computed via the FFT.

3.6.1 Simulation study

Examples will be provided to illustrate the method. In all the examples, the gain and phase margins are set to $A_m = 3$ and $\phi_m = \frac{\pi}{3}$(or 60 degrees) respectively. The default value of $\alpha = 1$ is also used. Following the previous discussion, step iii) in the tuning procedure may be dispensed as a solution to (3.24) is guaranteed under this set of gain and phase margin specification. Step responses using the Exact-GPM (PID) method are plotted in solid lines, whereas those using GPM method in dashed lines in the figures that follow.

Example 3.5:

Consider

$$G_p(s) = \frac{1}{(s+1)(s+3)}e^{-s}.$$

The phase crossover frequency of the process is $\omega_c = 1.626$ rad/s, and $G_p(j\omega_c) = -0.154$. Thus $\omega_p = 1.626$ rad/s. It follows from (3.19) that

$$K_p = Re\left[\frac{-1}{3 \times (-0.154)}\right] = 2.171.$$

The search is initiated from $\omega = \omega_c$ towards $\omega = 0$ to locate the frequency $\omega_g = 0.535$ rad/s which satisfies (3.24), i.e.

$$Re\left[\frac{-cos(60^o) - jsin(60^o)}{G_p(j0.535)}\right] = 2.171 = Re\left[\frac{-1}{3G_p(j1.626)}\right].$$

The intermediate values X_p and X_g are readily obtained to be 0 and -2.689. The values of K_i and K_d are then calculated using (3.20) and (3.21) as:

$$K_i = [0 - (-2.689) \times 1.626]\left(\frac{1.626}{0.535} - \frac{0.535}{1.626}\right)^{-1} = 1.614,$$

$$K_d = \left[0 - \frac{(-2.689)}{1.626}\right]\left(\frac{1.626}{0.535} - \frac{0.535}{1.626}\right)^{-1} = 0.611.$$

The designed PID controller is hence given by:

$$G_c(s) = 2.171 + \frac{1.614}{s} + 0.611s.$$

The step response of the resultant system is shown in Fig. 3.8. The gain and phase margins achieved are $A_m = 3.0$ and $\phi_m = 1.047$ (or 60.0 degrees) respectively. The PID parameters using GPM method are given by:

$$G_c(s) = 2.278 + \frac{1.712}{s} + 0.758s,$$

with gain and phase margins of $A_m = 3.0$ and $\phi_m = 1.032$ (or 59.1 degrees). Since GPM method nearly achieves the given margins, both methods yield similar performance as expected.

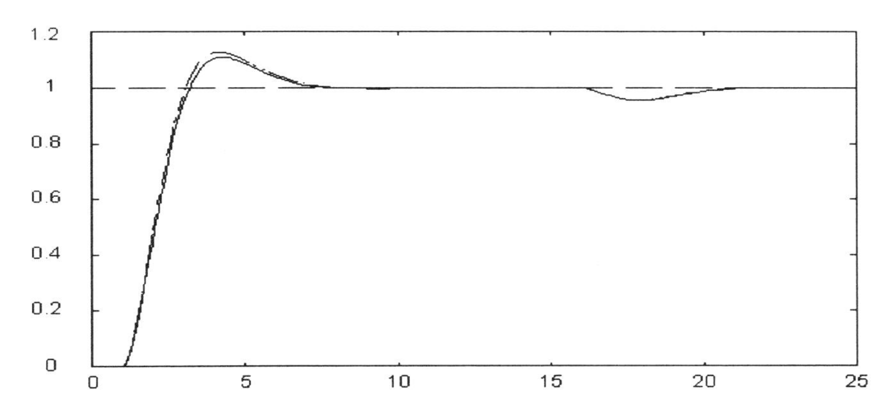

Fig. 3.8. Control performance with $G_p(s) = \frac{1}{(s+1)(s+3)}e^{-s}$ (solid:- Exact-GPM, dashed:- GPM): Example 3.5

Example 3.6:

Consider a second-order oscillatory process:

$$G_p(s) = \frac{1}{(s^2 + 2s + 3)}e^{-2s}.$$

It follows from the tuning procedure that the controller is,

$$G_c(s) = 0.947 + \frac{0.968}{s} + 0.775s.$$

Gain and phase margins of 3.0 and 1.047 (or 60.0 degrees) are achieved. The step response is shown in Fig. 3.9. Satisfactory response is obtained. The GPM method has no solution in this case.

Example 3.7:

Consider a high-order oscillatory process:

$$G_p(s) = \frac{1}{(s^2 + s + 5)(s + 1)}e^{-2s}.$$

The controller designed using the Exact-GPM method is given by

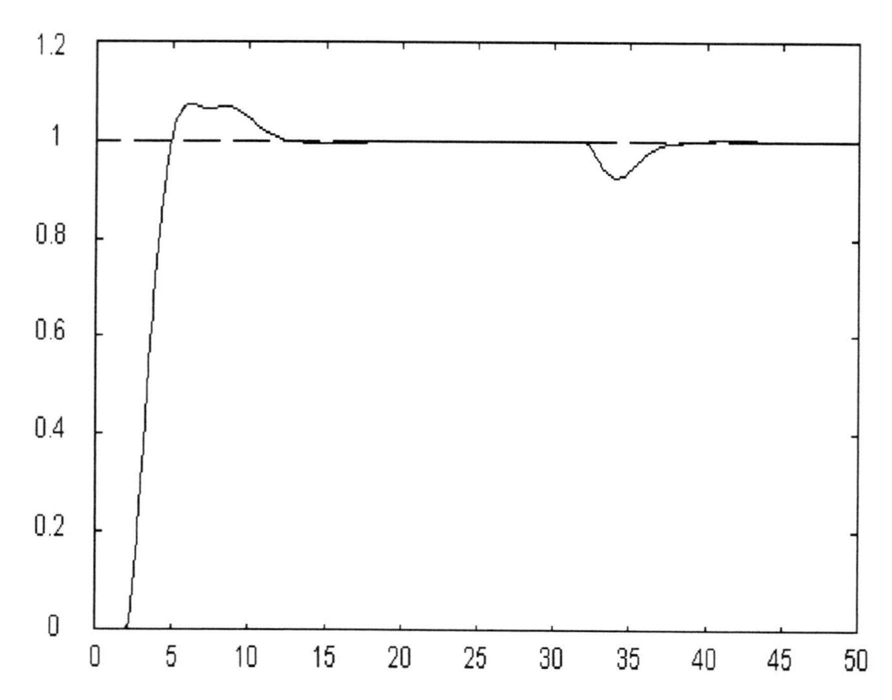

Fig. 3.9. Control performance with $G_p(s) = \frac{1}{(s^2+2s+3)}e^{-2s}$: Example 3.6

$$G_c(s) = 1.950 + \frac{1.478}{s} + 1.372s.$$

The resulting gain and phase margins are 3.0 and 1.047 (or 60.0 degrees). The step response is shown in Fig. 3.10. The controller obtained using GPM method is given by:

$$G_c(s) = 0.914 + \frac{1.154}{s} + 0.181s.$$

The corresponding gain and phase margins achieved are $A_m = 3.0$ and $\phi_m = 1.023$ (or 58.6 degrees) respectively. The Exact-GPM method gives rise to a significantly faster response to step change and disturbance.

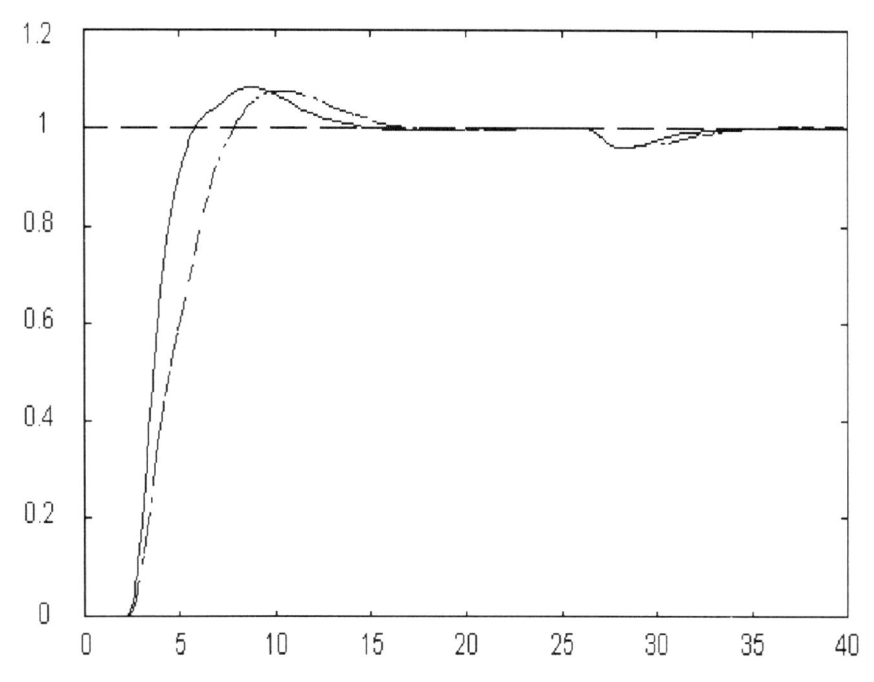

Fig. 3.10. Control performance with $G_p(s) = \frac{1}{(s^2+s+5)(s+1)}e^{-2s}$ (solid:- Exact-GPM, dashed:- GPM): Example 3.7

3.7 Linear Quadratic Control Design

Linear Quadratic Regulator (LQR) design technique is well known in modern optimal control theory and it has been widely used in many applications (Lewis et.al., 1995). It has a very nice robustness property, i.e., if the process is of single-input and single-output, then the control system has at least a phase margin of 60 degree and a gain margin of infinity. This attractive property appeals to the practitioners. Thus, the LQR theory has received considerable attention since 1950s. In the context of optimal PID tuning, typical performance indices are the integral of squared error and time weighted error. With this kind of performance criterions, the integral of squared error (squared time weighted error) is calculated using "Astrom's integral algorithm" recursively if the process transfer function is known (Zhuang et.al., 1993). The Pade approximation is used to replace the time-delay and then obtain the optimal PID controller for a first-order plus time-delay process. It is however noted that the Pade approximation may be inadequate for large normalized time-delay. It is also noticed that an analytical tuning formula cannot be obtained via this optimization. The computational procedure to minimize the performance criterion is complicated and thus unsuitable for on-line applications.

In this section, the LQR approach is employed to develop an optimal PI/PID controller tuning algorithm for the low-order plus time-delay model. A new criterion for selection of the and matrices is used which will lead to the desired natural frequency and damping ratio of the closed-loop system. The examples with various dynamics are included to demonstrate the effectiveness of the tuning algorithms and show significant improvement over some existing best PID tuning methods. Finally, the robustness property of the tuning algorithms is analyzed, and it is shown that the LQR system is robustly stable for small modeling errors.

3.7.1 LQR solution for time-delay systems

Consider a linear process with time-delay described by:

$$\dot{x}(t) = Ax(t) + Bu(t - L), \tag{3.30}$$

and the control performance specification measured in terms of

$$J = \int_0^\infty [x^T(t)Qx(t) + u^T(t)R(t)]dt, \tag{3.31}$$

where A, B, C, Q and R are given matrices with proper dimensions, $Q \geq 0$ and $R > 0$, $u(t) = 0$, when $t < 0$. The LQR problem is to find the optimal control $u(t)$ such that J in (3.31) is minimized. The dynamical process (3.30) is first decomposed into two stages:

(I) when $0 \leq t < L$, $u(t - L) = 0$, there is no input signal to process (3.30) so that

$$\dot{x}(t) = Ax(t), \quad 0 \leq t < L, \tag{3.32}$$

(II) when $t \geq L$, the process has a possible non-zero input signal. In this stage, let $\hat{u}(t) = u(t - L)$, $t \geq L$, and it follows:

$$\dot{x}(t) = Ax(t) + B\hat{u}(t), \quad t \geq L. \tag{3.33}$$

Through this transformation, (3.32) and (3.33) are now both delay-free and the LQR result for delay-free process can then be applied. It is well known (Lewis et.al., 1995) that the LQR solution to process (3.33) is

$$\hat{u}(t) = -R^{-1}B^T Px(t), \quad t \geq L, \tag{3.34}$$

where P is the positive definite solution of the Riccati equation:

$$A^T + PA - PBR^{-1}B^T P + Q = 0 \tag{3.35}$$

Converting $\hat{u}$ in (3.34) back to $u(t)$, the LQR solution to the original process (1) with the index (3.31) is obtained as:

$$u(t) = \hat{u}(t + L) = -R^{-1}B^T Px(t + L), \quad t \geq 0. \tag{3.36}$$

One sees from (3.36) that although the control law $\hat{u}(t)$ given in (3.34) is in time horizon of $t \geq L$, the recovered $u(t)$ actually gives the control signal for process (1) in the whole time horizon of $t \geq 0$. $x(t+L)$ is not directly available at time t. By (3.32)-(3.34), however, it can be expressed by the transmission of $x(t)$ as

$$x(t + L) = e^{(A-BR^{-1}B^T P)t}x(L) = e^{(A-BR^{-1}B^T P)t}e^{A(L-t)}x(t), \tag{3.37}$$

when $0 \leq t < L$ and

$$x(t + L) = e^{(A-BR^{-1}B^T P)t}x(L) = e^{(A-BR^{-1}B^T P)L}x(t), \tag{3.38}$$

when $t \geq L$. If the matrix Q is factorized as $Q = H^T H$, the LQR solution to (3.30) and (3.31) can thus be summarized (Marshall, 1979) in the following theorem.

Theorem 3.1:

For the linear process (3.30) with time-delay, if (A, B) is controllable and (H, A) is observable, then the optimal control minimizing the criterion function (3.31) is given by:

$$u(t) = -R^{-1}B^T P e^{(A-BR^{-1}B^T P)t} e^{A(L-t)} x(t), \quad 0 \le t < L, \tag{3.39}$$

and

$$u(t) = -R^{-1}B^T P e^{(A-BR^{-1}B^T P)L} x(t), \quad t \ge L, \tag{3.40}$$

where P is the positive definite solution to (3.35). The resultant system is also stable.

One may see from (3.36) that the current control $u(t)$ is actually a feedback of the future state at time of $(t + L)$. It implies that the controller has the prediction capability and thus may improve the closed-loop performance compared with traditional LQR or PID design. It is also noticed that during the starting period of time $t < L$, the control law (3.39) is time varying and generates a relatively large gain required to speed up the response. When $t = L$, (3.39) coincides with (3.40) and thus the control law is continuous. After that the feedback gain becomes constant, as seen in (3.40).

The major criticisms on LQR design, especially from the process control community, are that all the state variables are usually not measurable and the selection of the weighting matrices Q and R is not clear in order to meet the closed-loop performance specifications, such as overshoot and setting time. The objective here is to develop an optimal PI/PID tuning algorithm via the above outlined LQR solution for most typical industrial processes such that these problems can be solved and LQR design becomes truly useful for practical applications in process control.

3.7.2 PI tuning for first-order modeling

In the process industry, a large class of processes has monotonic input-output transients whose transfer functions can be approximated (Luyben, 1990) by a first-order plus time-delay one:

$$\tilde{G}_p(s) = \frac{b}{s+a} e^{-Ls} \tag{3.41}$$

It should be noted that (3.41) is not the process itself but a model of it and is used only for purpose of controller design. The controller, once designed, should be applied to the process but not the model. A PI controller:

$$u(t) = K_p\left(e(t) + \frac{1}{T_i}\int e(t)dt\right) = K_p e(t) + K_i \int e(t)dt \qquad (3.42)$$

is adequate for such a kind of processes (Astrom et al., 1988). In this section, an optimal PI tuning algorithm will be derived via the LQR approach of the last section and the closed formulae for selecting and in terms of the closed-loop specifications.

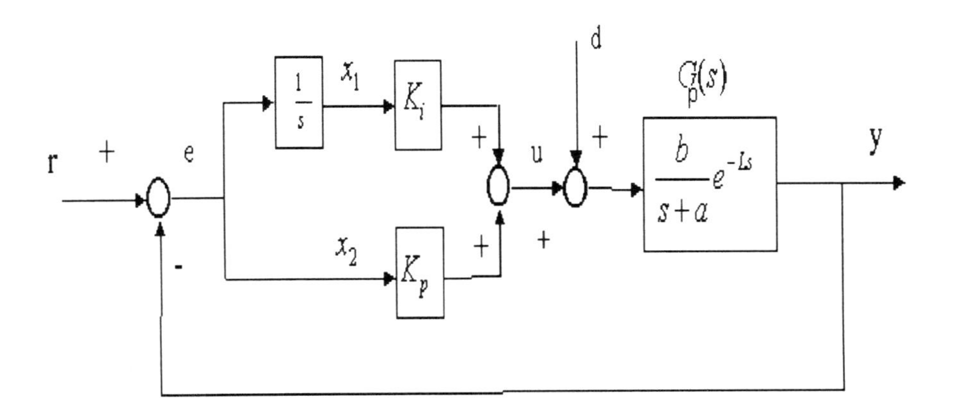

Fig. 3.11. Feedback control system

Consider a unity output feedback system shown in Fig. 3.11. In the case of feedback design, the external set-point does not affect the result and reference may be set at $r = 0$. It then follows from Fig. 3.11 that $(s+a)e = -be^{-Ls}u$, which is equivalent to the time domain equation:

$$\dot{e} = -ae - bu(t-L). \qquad (3.43)$$

The identity follows:

$$\frac{d}{dt}\int_0^t e(t)dt = e.$$

Let $x_1 = \int_0^t e(t)dt$ and $x_2 = e$ such that $x = [x_1 x_2]^T$. Then (3.43) can be put into the equivalent equation:

$$\dot{x}(t) = \begin{bmatrix} 0 & 1 \\ 0 & -a \end{bmatrix} x + \begin{bmatrix} 0 \\ -b \end{bmatrix} u(t-L), \tag{3.44}$$

It should be emphasized that both variables are available (Fig. 3.11) and the state feedback of K_x is simply $(K_i \int_0^t edt + K_p e)$, i.e., PI control. As a result, the state feedback gain to be derived using LQR design will give the required PI parameters.

In order to find the explicit expressions for and for ease of use, comparing (3.44) with (3.30) yields $A = \begin{bmatrix} 0 & 1 \\ 0 & -a \end{bmatrix}$ and $B = \begin{bmatrix} 0 \\ -b \end{bmatrix}$. Let $Q = \begin{bmatrix} q_1 & 0 \\ 0 & q_2 \end{bmatrix}$. Substituting $P = \begin{bmatrix} P_{11} & P_{12} \\ P_{21} & P_{22} \end{bmatrix}$ into the Riccati Equation (3.35) yields:

$$\begin{bmatrix} 0 & 1 \\ 0 & -a \end{bmatrix} \begin{bmatrix} P_{11} & P_{12} \\ P_{21} & P_{22} \end{bmatrix} + \begin{bmatrix} P_{11} & P_{12} \\ P_{21} & P_{22} \end{bmatrix} \begin{bmatrix} 0 & 1 \\ 0 & -a \end{bmatrix} - \begin{bmatrix} P_{11} & P_{12} \\ P_{21} & P_{22} \end{bmatrix} \begin{bmatrix} 0 \\ -b \end{bmatrix} R^{-1} \begin{bmatrix} 0 & -b \end{bmatrix} \begin{bmatrix} P_{11} & P_{12} \\ P_{21} & P_{22} \end{bmatrix} + \begin{bmatrix} q_1 & 0 \\ 0 & q_2 \end{bmatrix} = 0. \tag{3.45}$$

The positive definite analytical solution is:

$$\begin{cases} p_{12} = \sqrt{q_1 R}/b, \\ p_{22} = (-Ra + \sqrt{R^2 a^2 + Rb^2(2p_{12} + q_2)})/b^2, \\ p_{11} = ap_{12} + R^{-1} b^2 p_{12} p_{22}. \end{cases} \tag{3.46}$$

Let

$$F = R^{-1} B^T P = R^{-1} \begin{bmatrix} 0 & -b \end{bmatrix} \begin{bmatrix} P_{11} & P_{12} \\ P_{21} & P_{22} \end{bmatrix} = -R^{-1} b \begin{bmatrix} p_{12} & p_{22} \end{bmatrix}, \tag{3.47}$$

and

$$A_c = A - BF = \begin{bmatrix} 0 & 1 \\ 0 & -a \end{bmatrix} + \begin{bmatrix} 0 \\ -b \end{bmatrix} R^{-1} b \begin{bmatrix} p_{12} \ p_{22} \end{bmatrix}$$
$$= \begin{bmatrix} 0 & 1 \\ -R^{-1}b^2 p_{12} & -\sqrt{a^2 + R^{-1}b^2(2p_{12} + q_2)} \end{bmatrix}. \tag{3.48}$$

The optimal controller in (3.39) and (3.40) then reduce to

$$\begin{cases} u(t) = -Fe^{A_c t} e^{A(L-t)} x(t), & 0 \le t < L, \\ u(t) = -Fe^{A_c L} x(t), & t \ge L. \end{cases} \tag{3.49}$$

Remark 3.1:

Note that given the A and B matrices of the process, the optimal controller in (3.49) depends only on the gain F in (3.47), or only on p_{12} and p_{22} in the solution (3.46) of Riccati equation. Now, if the matrix Q in (3.45) is replaced by a general form of $Q = \begin{bmatrix} q_1 & q_{12} \\ q_{12} & q_2 \end{bmatrix}$, then the positive definite analytical solution is

$$p_{12} = \frac{1}{b}\sqrt{q_1 R},$$

$$p_{22} = \frac{-Ra + \sqrt{R^2a^2 + Rb^2(2p_{12} + q_2)}}{b^2},$$

and

$$p_{11} = ap_{12} + R^{-1}b^2 p_{12} p_{22} - q_{12}.$$

Note that p_{12} and p_{22} remain the same as those in (3.46) though the matrix is non-diagonal. This shows that choosing with a diagonal form will not lose the generality in the PI optimal controller design via the LQR approach for the case in Fig. 3.11.

To obtain the feedback gains in (3.49) explicitly, one needs to calculate $exp(A_c t)$ and $exp(A(L-t))$. It follows from the inverse Laplace transformation that:

$$e^{A(L-t)} = \ell^{-1}(sI - A)^{-1}|_{(L-t)} = \begin{bmatrix} 1 & \frac{1-e^{-a(L-t)}}{a} \\ 0 & e^{-a(L-t)} \end{bmatrix}. \tag{3.50}$$

As for $e^{A_c t}$, let $\hat{a}_1 = \sqrt{a^2 + R^{-1}b^2(2p_{12} + q_2)}$, $\hat{a}_2 = R^{-1}b^2 p_{12}$, α_1 and α_2 be the roots of the equation $s^2 + \hat{a}_1 s + \hat{a}_2 = 0$, i.e., $\alpha_1 = \frac{-\hat{a}_1 + \sqrt{\hat{a}_1^2 - 4\hat{a}_2}}{2}$ and $\alpha_2 = \frac{-\hat{a}_1 - \sqrt{\hat{a}_1^2 - 4\hat{a}_2}}{2}$. It follows that

$$e^{A_c t} = \ell^{-1}(sI - A)^{-1} = \begin{bmatrix} f_{11} \; f_{12} \\ f_{21} \; f_{22} \end{bmatrix}, \tag{3.51}$$

where

$$f_{11}(t) = \frac{1}{\alpha_1 - \alpha_2}[(\alpha_1 + \hat{a}_1)e^{\alpha_1 t} - (\alpha_2 + \hat{a}_1)e^{\alpha_2 t}],$$

$$f_{12} = \frac{1}{\alpha_1 - \alpha_2}[e^{\alpha_1 t} - e^{\alpha_2 t}],$$

$$f_{21} = \frac{-\hat{a}_2}{\alpha_1 - \alpha_2}[e^{\alpha_1 t} - e^{\alpha_2 t}],$$

$$f_{22} = \frac{1}{\alpha_1 - \alpha_2}[\alpha_1 e^{\alpha_1 t} - \alpha_2 e^{\alpha_2 t}].$$

Recall that $u = K_x = [K_i \; K_p][\int_0^t e \, dt \; e]^T$. Substituting (3.47), (3.48), (3.50) and (3.51) into (3.49) gives the explicit expressions for the PI parameters.

Theorem 3.2:

The LQR optimal control for process (3.41) with state equation (3.44) is given in the form of a PI controller (3.42), where for $0 \leq t < L$,

$$\begin{aligned} K_i(t) &= R^{-1}b[p_{12}f_{11}(t) + p_{22}f_{21}(t)], \\ K_p(t) &= R^{-1}b\frac{1}{a}p_{12}f_{11}(t) + \frac{1}{a}p_{22}f_{21}(t) + [p_{12}f_{12}(t) - \frac{1}{a}p_{12}f_{11}(t) \\ &\quad + p_{22}f_{22}(t) - \frac{1}{a}p_{22}f_{21}(t)]e^{-a(L-t)}; \end{aligned} \tag{3.52}$$

and for $t \geq L$,

$$\begin{aligned} K_i(t) &= R^{-1}[p_{12}f_{11}(L) + p_{22}f_{21}(L)], \\ K_p(t) &= R^{-1}[p_{12}f_{12}(L) + p_{22}f_{22}(L)], \end{aligned} \tag{3.53}$$

where constants p_{12} and p_{22} are given in (3.46), $f_{ij}(t)$, $i = 1, 2$; $j = 1, 2$, are given in (3.51), q_1, q_2 and R are tuning parameters.

In an ordinary LQR design, the selection of the Q and R matrices is quite technical and affects the system performance significantly. In order to overcome this difficulty, a direct relationship will be derived between q_1 and q_2, and the damping ratio ξ and natural frequency ω_n of the closed-loop system.

Theorem 3.3:

When $t \geq L$, the damping ratio ξ and the natural frequency ω_n of the LQR optimal closed-loop system in (3.42) and (3,44) is:

$$\begin{cases} \omega_n = \sqrt{R^{-1}b\sqrt{q_1 R}}, \\ \xi = \frac{\sqrt{a^2+R^{-1}b(2\sqrt{q_1 R}+q_2 b)}}{2\sqrt{R^{-1}b\sqrt{q_1 R}}}. \end{cases} \tag{3.54}$$

Equivalently, in order to have the desired ξ and ω_n, q_1 and q_2 should be chosen as:

$$\begin{cases} q_1 = \frac{\omega_n^4 R}{b^2}, \\ q_2 = \frac{[(4\xi^2-2)\omega_n^2-a^2]R}{b^2}. \end{cases} \tag{3.55}$$

Proof:

When $t \geq L$, the closed-loop system becomes:

$$\dot{x} = A_c x = \begin{bmatrix} 0 & 1 \\ -R^{-1}b\sqrt{q_1 r} & -\sqrt{a^2 + R^{-1}b(2\sqrt{q_1 r} + q_2 b)} \end{bmatrix},$$

and the characteristic equation is:

$$\Delta = s(s + \sqrt{a^2 + R^{-1}b(2\sqrt{q_1 r} + q_2 b)}) + R^{-1}b\sqrt{q_1 R}.$$

It thus has

$$\begin{cases} \omega_n^2 = R^{-1}b\sqrt{q_1 R}, \\ 2\xi\omega_n = \sqrt{a^2 + R^{-1}b(2\sqrt{q_1 R} + q_2 b)}. \end{cases}$$

The theorem follows directly.

Remark 3.2:

For the system (3.44) with $Q = diag\{q_1, q_2\}$ and q_1 and q_2 chosen according to (3.55), the performance index (3.31) becomes:

$$J = R\left[\int_0^\infty \left[\frac{\omega_n^4}{b^2}\left(\int_0^t e(t)\,dt\right)^2 + \frac{(4\xi^2-2)\omega_n^2 - a^2}{b^2}e(t)^2 + u(t)^2\right]dt\right],$$

i.e., J is proportional to R. This implies that R makes no sense in the design of controller gain F in (3.47) and thus it can always be chosen as $R = 1$ when Theorem 3.3 is applied.

In view of the above development, an optimal PI tuning algorithm for process (3.31) can be summarized as follows for ease of reference.

Optimal PI tuning algorithm. Initialization: Obtain a, b, L and set $R = 1$.
Step 1. Choose the closed-loop ω_n and ξ.
Step 2. Calculate q_1 and q_2 from (3.45).
Step 3. Calculate p_{12} and p_{22} from (3.46), A_c from (3.48) and $e^{A_c t}$ from (3.51).
Step 4. Calculate the PI parameters from (3.52) and (3.53).

Remark 3.3:

In the LQR algorithm, ξ and ω_n are the only user-specified parameters. From extensive trial tests, choosing $\xi \in [0.7, 1.0]$ and $\omega_n L \in [1.0, 1.5]$ would give a satisfactory result. Normally, the defaults of $\xi = 0.71$ and $\omega_n L = 1.3$ can be used. For a better performance, a finer tuning procedure may be used.

3.7.3 Simulation study

To illustrate the LQR tuning algorithm, different processes are considered in this section. A model for the process may be obtained by process-model matching at two frequencies (Luyben, 1990) or by least square fitting between process and model frequency responses (Wang et.al., 1997). The later is used for the simulation. For comparison, the PI controller tuning by Ho's gain-phase margin method (GPM) (Ho et.al, 1995b) is again employed. For a fair comparison, the same processes with the same identified models will be used in the simulations. Typical gain margin of $A_m = 3$ and phase margin of $\phi_m = 45^0$ are used for Ho's algorithm throughout the examples and the corresponding closed-loop responses are shown by dotted lines. Solid lines in the simulation below reflects the response from the LQR optimal PI control system. Within the time duration of the simulation, a disturbance of $d = 0.2$ is introduced so that the disturbance attenuation property of the designed controller is also illustrated.

Example 3.8:

Consider a high vacuum distillation column (Hang et al., 1993). The transfer function between the viscosity and the reflux flow is given by:

$$G_p(s) = \frac{0.57}{(1+8.60s)^2}e^{-18.70s},$$

and the model is identified as:

$$\tilde{G}_p(s) = \frac{0.57}{(1+12.72s)^2}e^{-23.2s}.$$

Choosing $\omega_n L = 1.3$ and $\xi = 0.71$, the PI controller's parameters are obtained as:

$$\begin{aligned} K_i(t) &= [0.0701cos(0.0395t) + 0.0690sin(0.0395t)]e^{-0.0398t}, \\ K_p(t) &= [0.1417sin(0.0395t) - 0.1404cos(0.0395t)]e^{0.0388t} \\ &\quad +[0.8913cos(0.0395t) + 0.8771sin(0.0395t)]e^{-0.0398t}, \end{aligned}$$

for $0 \le t < 23.2$ and $K_i = 0.0387$, $K_p = 0.5581$ for $t \ge 23.2$. The control signal and the closed-loop performance are shown in Fig. 3.12. The results show that the LQR method gives a perfect response and significantly outperforms that of the GPM method.

Example 3.9:

Consider the high-order process:

$$G_p(s) = \frac{1}{(s+1)^n},$$

with n=10 and n=20. The resultant models are respectively:

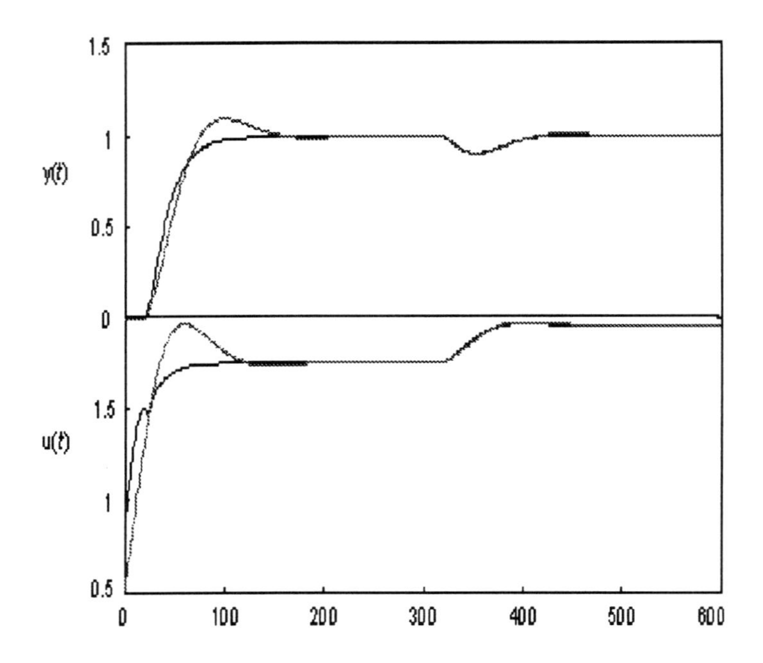

Fig. 3.12. Control performance with $G_p(s) = \frac{0.57}{(1+8.60s)}e^{-18.70s}$ (overshoot:- GPM, without overshoot:- LQR): Example 3.8

$$\tilde{G}_p(s) = \frac{1}{1+2.72s}e^{-7.69s}, n = 10,$$

$$\tilde{G}_p(s) = \frac{1}{1+4.95s}e^{-15.67s}, n = 20.$$

The closed-loop responses in Fig. 3.13 show that the LQR method provides a much better performance than GPM method (GPM).

Example 3.10:

The algorithm is also applied to the non-minimum-phase process:

$$G_p(s) = \frac{1-\alpha s}{(1+s)^3},$$

with $\alpha = 1$ and $\alpha = 1.5$. The closed-loop performance shown in Fig. 3.14 exhibits a great improvement with the LQR method.

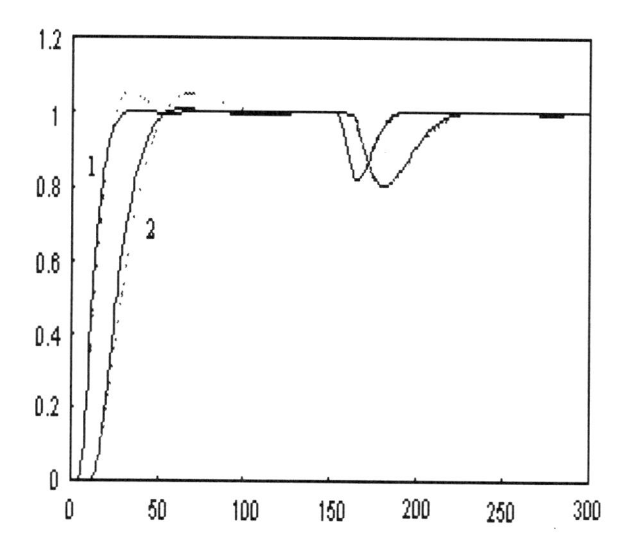

Fig. 3.13. Control performance with $G_p(s) = \frac{1}{(s+1)^n}$; $n = 10(1)$ and $n = 20(2)$ (solid:- LQR, dashed:- GPM): Example 3.9

The simulations from the above examples show that the LQR-based PI tuning algorithm gives a much better closed-loop performance over some well known PI tuning methods. One also sees that the control signal given by the LQR method is larger than that of GPM method when $t < L$, and it leads to the faster setpoint response. But thereafter the gains decrease and the overall control signal amplitude is no large than that for ordinary PI tuning. For these three examples, the actual processes, their models, control specifications and PI parameters are listed in Table 3.1 for ease of reference.

3.7.4 Extension to second-order modeling

PI control is sometimes inadequate when the process dynamics is not essentially first-order (Astrom and Hagglund, 1988). In this section, the PID tuning formula will be derived for the second-order plus time-delay model instead of the first-order plus time-delay model. Consider a second-order process model given by:

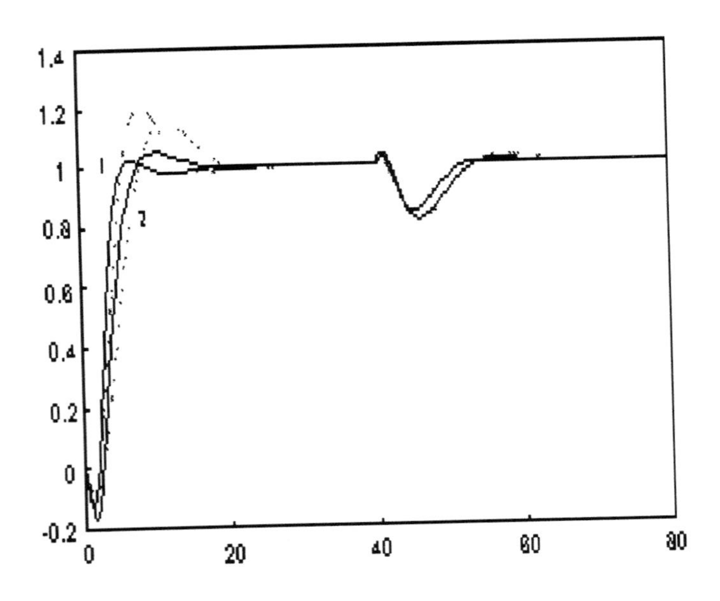

Fig. 3.14. Control performance with $G_p(s) = \frac{1-\alpha s}{(1+s)^3}$; $\alpha = 1(1)$ and $\alpha = 1.5(2)$: Example 3.10

$$\tilde{G}_p(s) = \frac{b}{(s+a)(s+a_1)} e^{-Ls}, \; a_1 \geq a. \tag{3.56}$$

A PID controller is written in the form:

$$G_c(s) = \left(K_p + K_i \frac{1}{s}\right)(s + K_d). \tag{3.57}$$

Then, from (3.56) and (3.57), the open-loop transfer function is given by:

$$G_c(s)G_p(s) = (K_p + K_i \frac{1}{s}) \frac{b(K_d + s)}{(s+a)(s+a_1)} e^{-Ls} \tag{3.58}$$

Table 3.1. PI controller parameters for different processes

Process	Model	Tuning	K_p	K_i	$\omega_n L(A_m)$	$\xi(\phi_m)$
$\frac{0.57e^{-18.70s}}{(1+8.60s)^2}$	$\frac{0.57e^{-23.2s}}{1+12.72s}$	LQR	0.5581	0.0387	1.3	0.71
		GPM	0.4719	0.0408	3	45
$\frac{1}{(1+s)^{10}}$	$\frac{e^{-7.69s}}{1+2.72s}$	LQR	0.2487	0.070	1.4	0.71
		GPM	0.1735	0.0680	3	45
$\frac{1}{(1+s)^{20}}$	$\frac{e^{-15.67s}}{1+4.95s}$	LQR	0.2335	0.0349	1.4	0.71
		GPM	0.1550	0.0331	3	45
$\frac{1-s}{(1+s)^3}$	$\frac{e^{-2.25s}}{1+1.61s}$	LQR	0.3806	0.2240	1.2	0.71
		GPM	0.3511	0.2468	3	45
$\frac{1-1.5s}{(1+s)^3}$	$\frac{e^{-2.89s}}{1+1.01s}$	LQR	0.2574	0.1850	1.5	0.71
		GPM	0.1715	0.1807	3	45

For ease of control design, one may choose $K_d = a_1$ to cancel the larger process pole. (3.58) is then reduced to:

$$G_c(s)G_p(s) = \left(K_p + K_i\frac{1}{s}\right)\frac{b}{(s+a)}e^{-Ls}. \tag{3.59}$$

Note that (3.59) now gives the same open-loop transfer function as with the first-order plus dead-time model $G_p(s)$ given in (3.41) and a PI controller in (3.42). Therefore, a PID controller can be simply tuned with $K_d = a_1$ and the PI parameters given as in the last section.

Similarly to the first-order plus dead-time model, a second-order plus time-delay model in (3.56) can be identified by various identification methods (Luyben, 1990). In Example 3.11, the least square fitting method is employed to identify the model and then the LQR method is applied. GPM tuning method for PID (Ho et.al. 1995a) is adopted again for comparison.

Example 3.11:

Consider a non-minimum phase process:

$$G_p(s) = \frac{1-s}{(1+s)^2(2+s)},$$

and the model is obtained as:

$$\tilde{G}_p(s) = \frac{e^{-1.64s}}{(s+1)(s+2)}.$$

Choosing $\omega_n L = 1.3$ and $\xi = 0.8$, the PID parameters are obtained as $K_p = 0.6138$, $K_i = 0.5561$ and $K_d = 1$ for $t \geq 1.64$. The simulation result in Fig. 3.15 shows a significant improvement of the LQR method over the GPM with the gain and phase margins specified as $A_m = 3$ and $\phi_m = 45^o$.

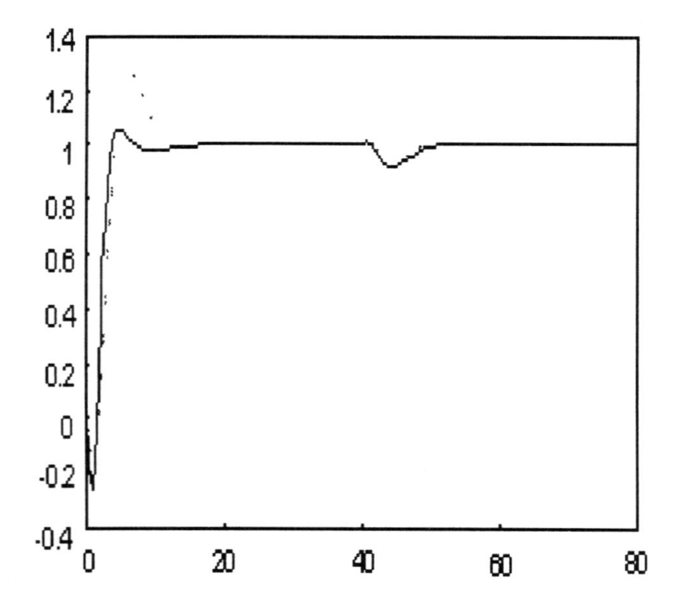

Fig. 3.15. Control performance with $G_p(s) = \frac{1-s}{(1+s)^2(s+2)}$ (solid:- LQR, dashed:- GPM): Example 3.11

3.7.5 Robustness analysis

One of the most attractive properties of LQR design for delay-free processes is the robustness of its closed-loop system, which is usually wanted in practical applications. If the process is of single-input and single-output, the resultant LQR system has at least a phase margin of 60 degree and a gain margin of infinity. Unfortunately, it is found that this property cannot be generalized to the time-delay case. The extension is only possible for special systems.

The stability of the controlled system in (3.30) and (3.40) is now considered with the real parameters perturbed to A_r, B_r and L_r. Without loss of generality, the robustness issue is considered when $t \geq max\{L, L_r\}$, because during a finite time interval, the system cannot go into infinity. If the control law of (3.40) is applied to the process (3.30), the resultant closed-loop becomes:

$$\dot{x}_r = A_r X_r - B_r R^{-1} B^T P e^{A_c L} X_r(t - L_r), \tag{3.60}$$

where $A_c = A - BR^{-1}B^T P$. Let matrix $\hat{A}_c$ be the solution of equation:

$$\hat{A}_c = A_r - B_r R^{-1} B^T P e^{A_c L} e^{-\hat{A}_c L_r}, \tag{3.61}$$

then the following theorem follows.

Theorem 3.4:

The perturbed system with real process parameters A_r, B_r and L_r remains stable if all eigenvalues of $\hat{A}_c$ given by (3.61) lie in the open left half of the complex plane.

Proof:

Substituting (3.61) into (3.60) yields

$$\begin{aligned}\dot{x}_r(t) &= [\hat{A}_c + B_r R^{-1} B^T P e^{A_c L} e^{-\hat{A}_c L_r}] x_r(t) \\ &\quad - B_r R^{-1} B^T P e^{A_c L} x_r(t - L_r).\end{aligned} \tag{3.62}$$

Consider the dynamical equation:

$$\dot{x}(t) = \hat{A}_c x(t). \tag{3.63}$$

The state transition satisfies:

$$x(t - L_r) = e^{-\hat{A}_c L_r} x(t).$$

Observe that the two differential equations (3.62) and (3.63) coincide. Thus, (3.62) is stable if (3.63) is stable, i.e., all the eigenvalues of $\hat{A}_c$ given by (3.63) lie in the open left half of the complex plane. The proof is completed.

Now, (3.61) is considered in the special case of the PI controller (3.42) for process (3.41) with real parameters $A_r = \begin{bmatrix} 0 & 1 \\ 0 & -a_r \end{bmatrix}$, $B_r = \begin{bmatrix} 0 \\ -b_r \end{bmatrix}$ and L_r. Let

$$e^{A_c L} e^{-\hat{A}_c L_r} = I + \Delta = \begin{bmatrix} 1 + \Delta_{11} & \Delta_{12} \\ \Delta_{21} & 1 + \Delta_{22} \end{bmatrix}. \tag{3.64}$$

Note that in the case of no modeling error in process (3.41), i.e., $A_r = A$, $B_r = B$ and $L = L_r$, then $\hat{A}_c = A_c$ will be the solution of (3.61), or $e^{A_c L} e^{-\hat{A}_c L_r} = I$ and $\Delta_{ij} = 0, i, j = 1, 2$. In general, however, the solution of (3.61 is continuous with respect to A_r, B_r and L_r. In other words, for any $\varepsilon > 0$, there exists a $\delta > 0$ such that if $\| A_r - A \| < \delta$, $\| B_r - B \| < \delta$ and $|L_r - L| < \delta$, then $|\Delta_{ij}| < \varepsilon$, i,j=1,2. The following proposition further shows that when the process perturbation is small enough, the control system will remain stable.

Proposition 3.1:

If the process parameter perturbations are small enough, the control system in (4.41) and (4.42) is robustly stable.

Proof:

Substituting A_r, B_r, L_r and (3.64) into (3.61) gives

$$\hat{A}_c = \begin{array}{l} \begin{bmatrix} 0 & 1 \\ 0 & -a_r \end{bmatrix} - \begin{bmatrix} 0 \\ R^{-1}b_r b \end{bmatrix} [p_{12} \ \ p_{22}](I + \Delta) \\ = \begin{bmatrix} 0 & -R^{-1}b_r b p_{12} - R^{-1}b_r b(p_{12}\Delta_{11} + p_{22}\Delta_{21}) \\ 1 & -a_r - R^{-1}b_r b(p_{22} + p_{12}\Delta_{12} + p_{22}\Delta_{22}) \end{bmatrix}^T \end{array}$$

The eigenvalues of $\hat{A}_c$ lie in the open left half of the complex plane if

$$\begin{cases} -R^{-1}b_r b p_{12} - R^{-1}b_r b(p_{12}\Delta_{11} + p_{22}\Delta_{21}) < 0, \\ -a_r - R^{-1}b_r b(p_{22} + p_{12}\Delta_{12} + p_{22}\Delta_{22}) < 0, \end{cases}$$

or

$$\begin{cases} \Delta_{11} + \frac{p_{22}}{p_{12}}\Delta_{21} > -1, \\ \Delta_{22} + \frac{p_{12}}{p_{22}}\Delta_{12} > -1. \end{cases} \tag{3.65}$$

Note from (3.50) that p_{12} and p_{22} are positive real numbers, and (3.65) will hold true if a ε is chosen such that $\varepsilon < min\{0.5, \frac{p_{12}}{2p_{22}}, \frac{p_{22}}{2p_{12}}\}$. Thus, for such an ε and $|\Delta_{ij}| < \varepsilon$, i,j=1,2, there exists a corresponding $\delta > 0$ such that if the parameter perturbations in the process satisfy $\| A_r - A \| < \delta$, $\| B_r - B \| < \delta$ and $|L_r - L| < \delta$, the resultant closed-loop system remains stable. The proposition is proven.

3.8 Composite PI-Adaptive Control Design

While the PID controller is widely applicable, for general nonlinear systems, the performance associated with PI control remains largely restricted and the gain scheduling strategy is often used, which view the nonlinear system as a series of linear ones, each valid around a certain operating point.

In this section, an adaptive PI control scheme is presented for a class of nonlinear systems. First, the nonlinear system is viewed as a first-order dominant linear model with an uncertainty that is possibly nonlinear and time-varying. Secondly, the PI control is applied to stabilize the dominant model and achieve a satisfactory system response. Thirdly, the Radis Basis Function (RBF) is employed to design an adaptive compensator to deal with the nonlinear uncertainty of the system. The composite control can guarantee the system to converge towards the desired equilibrium state of the system, or a residual sphere around the equilibrium. Two practical examples are provided to illustrate the effectiveness of the PI-adaptive control scheme.

The essence of the motivation behind the PI-adaptive control scheme is to retain the widely accepted PI control structure as the foundation of the composite control structure and yet to achieve a performance level outside of the capability of conventional PI. From a practical viewpoint, this composite scheme is desirable since the operator will continue to tune only the PI portion of the controller, or to use an appropriate PID auto-tuning strategy. When the performance achieved still fall short of requirement, the RBF adaptation portion can be enabled to yield enhanced performance. The operation of the adaptive compensator essentially occurs behind the forefront of PI control and is oblivious to the operator.

3.8.1 Problem formulation

In the process control industry, a large class of nonlinear processes exhibit monotonic input-output transients which can be approximated (Luyben, 1990) as a first-order plus dead-time transfer function model described as:

$$\tilde{G}_p(s) = \frac{b}{s+a}e^{-Ls}. \tag{3.66}$$

For this class of processes, it is adequate (Astrom et al., 1993) to consider a PI controller described by

$$u(t) = K_p e(t) + K_i \int e(t)dt. \tag{3.67}$$

When $L > 0$, PI control may be designed based on a predictive approach such as that in Section 3.8. For illustration, here the case of $L = 0$ is considered. The linear model (3.66) can also be expressed as the form of the time domain:

$$\dot{y}(t) = -ay(t) + bu(t), \tag{3.68}$$

where $y(t)$ is the output and $u(t)$ is the control.

For processes exhibiting strong nonlinear and time-varying behavior, the above model alone may become inadequate especially when control performance requirements are stringent. For an example, the process is often augmented by extra nonlinear terms, $\bar{f}(y, u)$ caused by model imperfections and/or exogenous factors like incipient faults that actually perturb the nominal dynamics (3.68). Thus, the following uncertain system is considered :

$$\dot{y}(t) = -ay(t) + bu(t) + b[\bar{f}(y(t), u(t)) + \bar{d}(t)]. \tag{3.69}$$

It is assumed that $\bar{f}(y(t), u(t))$ is a smooth nonlinear function (unknown) and the disturbance signal $\bar{d}(t)$ (unknown) has a constant known upper bound value. This model represents a large and rich class of nonlinear systems.

In many cases, for a nonlinear system, a gain-scheduling strategy is used which requires a series of linear models around different operating points. The time and effort associated with the tuning and maintenance of such control schemes naturally increases tremendously. In this section, a different approach is experimented. PI control structure serves as the main foundation of the strategy with a fixed set of gains. Therefore, to the operator, the requirements for tuning and maintenance of the controller remain largely unchanged. However, to achieve a high level of control performance, an adaptive compensator based on a RBF is further incorporated to expand the capability of the conventional PI controller.

The problem is to consider tracking control for the system (3.69). For a given setpoint r_f (constant), the tracking error is

$$e(t) = r_f - y(t). \tag{3.70}$$

The desired control for the linear model is

$$u^* = \frac{a}{b} r_f. \tag{3.71}$$

Differentiating $e(t)$ and using (3.69), the system dynamics may be written as:

$$\dot{e}(t) = -ae(t) - bu(t) - b[f(e(t), \delta u(t)) + d(t)], \tag{3.72}$$

where $\delta u = u^* - u(t)$, $f(e(t), \delta u(t)) = \bar{f}(r_f - e(t), u^* - \delta u(t))$, and $d(t) = \bar{d}(t) - \frac{a}{b} r_f$.

Since

$$\frac{d}{dt} \int_0^t e(t)dt = e(t), \tag{3.73}$$

let the system's variables be $x_1 = \int_0^t e(t)dt$ and $x_2 = e(t)$ such that $x = [x_1 \; x_2]^T$. Then (3.72) can be put into the equivalent state equation:

$$\dot{x} = \begin{bmatrix} 0 & 1 \\ 0 & -a \end{bmatrix} x + \begin{bmatrix} 0 \\ -b \end{bmatrix} u(t) + \begin{bmatrix} 0 \\ -b \end{bmatrix} [f(e, \delta u) + d(t)]. \tag{3.74}$$

It should be emphasized that both state variables are available.

Consider the robust control:

$$u(t) = u_{pi} + u_a = Kx(t) - \hat{f}(e, \delta u) \tag{3.75}$$

where $\hat{f}(e, \delta u)$ is the estimated function of $f(e, \delta u)$. It is noted that the state feedback of $Kx(t)$ is simply $(K_i \int_0^t edt + K_p e)$, i.e., PI control. In what follows, the design of u_{pi} and u_a components will be investigated.

3.8.2 PI control based on first-order dominant model

In this subsection, explicit expressions for K_i and K_p based on simple user specifications will be derived. The nominal portion (without uncertainty) of the model is:

$$\dot{x} = \begin{bmatrix} 0 & 1 \\ 0 & -a \end{bmatrix} x + \begin{bmatrix} 0 \\ -b \end{bmatrix} u_{pi}, \tag{3.76}$$

where

$$u_{pi} = K_i x_1 + K_p x_2. \tag{3.77}$$

This is a standard PI control problem and much work has been done in this area. In principle, the existing PI tuning methods such as the gain and phase margin (GPM) method (Ho et al., 1995), dominant pole method (Astrom et al., 1993) etc., may all be employed. For this particular model used, a simple method for obtaining PI parameters associated with the LQR solution (Section 3.7) can be easily obtained.

The LQR problem requires a minimization of an index of control accuracy of the form:

$$J = \int_0^\infty (x^T Q x + \lambda u^2) dt, \tag{3.78}$$

where $Q \geq 0$ and $\lambda > 0$. It is well known that the LQR solution to (3.77) with the performance index (3.78) is given by:

$$u_{pi} = -\lambda^{-1} B^T P x(t), \tag{3.79}$$

where P is the positive definite solution of the Riccati equation

$$A^T P + PA - \lambda^{-1} PBB^T P + Q = 0, \tag{3.80}$$

and $Q = H^T H$.

One nice feature associated with LQR design is that under mild conditions, the resultant closed-loop system is always stable.

In order to implement the control law in the absence of knowledge of the bounds outside uncertainties, it is desirable to design an adaptive scheme for $\hat{f}(e, \delta u)$. This will be illustrated in the next subsection.

3.8.3 Nonlinear adaptive control

The design of the adaptive compensator for the nonlinear part of the system (3.74) is now considered. Since the nonlinear part is unknown, it is an objective of the paper to design a RBF compensator to capture the nonlinearities of the system. The main property of a RBF used here for estimation purposes is the function approximation property (Hornik et al.,1989; Fabri and Kadirkamanathan, 1996).

Let $f(\chi)$ be a smooth function from R^n to R^p. Then, given a compact $S \in R^n$ and a positive number ε_M, there exists a RBF system such that

$$f(\chi) = \sum_{i=0}^{m} W_i \phi_i(\chi) + \varepsilon, \tag{3.81}$$

where W_i is the representative value vector and $\phi_i(\chi)$ is the radial basis function with $||\varepsilon|| < \varepsilon_M$ for all $\chi \in S$.

Since $f(e, \delta u)$ is a nonlinear smooth function (unknown), it may be represented by RBF with constant 'ideal' weights $w_i, i = 1, 2, ..m$ and a sufficient number of basis functions $\phi(\cdot)$, i.e.

$$f(e, \delta u) = \sum_{i=0}^{m} w_i \phi_i(e, \delta u) + \epsilon(e, \delta u) = \sum_{i=0}^{m} \phi_i(e, \delta u) w_i + \epsilon(e, \delta u), \tag{3.82}$$

with $|\epsilon(e, \delta u)| \leq \epsilon_M$, where $\phi_i(e, \delta u)$ is given by:

$$\phi_i(e, \delta u) = exp(-\frac{||\kappa_e - c_i||^2}{2\sigma_i^2}) / \sum_{j=0}^{m} exp(-\frac{||\kappa_e - c_j||^2}{2\sigma_j^2}), \tag{3.83}$$

where $\kappa_e = [e, \delta u]^T$, c_i is 2-dimensional vector representing the center of the ith basis function, and σ_j is the variance representing the spread of the basis function.

Let the RBF functional estimates for $f(e, \delta u)$ be given by:

$$\hat{f}(e, \delta u) = \sum_{i=0}^{m} \phi_i(e, \delta u)\hat{w}_i, \tag{3.84}$$

where $\hat{w}_i$ are estimates of the ideal RBF weights that are provided by the following weight-tuning algorithm:

$$\dot{\hat{w}}_i = r_1 x^T P B \phi_i - r_2 \hat{w}_i, \tag{3.85}$$

where $r_1, r_2 > 0$, and P is the solution of (3.80).

It will be next demonstrated that the state x and weights remain bounded. The following definition is necessary for the illustration.

Definition 3.1:

Given $x \in R^n$ and a nonlinear function $h(x,t) : R^n \times R \to R^n$, the differential equation

$$\dot{x} = h(x,t), t_0 \leq t, x(t_0) = x_0, \tag{3.86}$$

has a differential solution $x(t)$ if $h(x,t)$ is continuous in $x(t)$ and t. The solution $x(t)$ is said to be uniformly ultimately bounded (UUB) if there exists a compact set $U \subset R^n$ such that, for all $x(t_0) = x_0 \in U$, there exists a δ>0 and a number $T(\delta, x_0)$ such that $||x(t)||<\delta$ for all $t \geq t_0 + T$ (Lewis et al.,1993).

Theorem 3.5:

Consider the case where the controller (3.75),(3.77) and (3.85) is applied to the system (3.74). The states and the RBF estimation errors are UUB.

Proof:

The differential equation (3.74) can be written as (upon applying the control (3.75),(3.77) and (3.85)):

$$\begin{aligned}
\dot{x} &= Ax + BKx - B\hat{f}(e,\delta u) + B[f(e,\delta u) + d(t)],\\
&= (A+BK)x - B\sum_{i=0}^{m}\phi_i(e)\hat{w}_i + B[\sum_{i=0}^{m}\phi_i(e)w_i + \epsilon(e) + d(t)],\\
&= (A - \lambda^{-1}BB^TP)x + B[\sum_{i=0}^{m}\phi_i(e)\tilde{w}_i + \epsilon(e) + d(t)]. \qquad (3.87)
\end{aligned}$$

Now consider the following Lyapunov function candidate:

$$V(x,\tilde{w}) = x^TPx + \frac{1}{r_1}\sum_{i=0}^{m}\tilde{w}^2. \qquad (3.88)$$

Taking the time derivative of v along the solution of (3.87), it can be shown that:

$$\begin{aligned}
\dot{V} &= x^T[(A-\lambda^{-1}BB^TP)^TP + P(A-\lambda^{-1}BB^TP)]x\\
&+ 2x^TPB\sum_{i=0}^{m}\phi_i\tilde{w}_i + 2x^TPB[\epsilon(e)+d(t)] + \frac{2}{r_1}\sum_{i=0}^{m}\tilde{w}_i\dot{\tilde{w}}_i,\\
&= -\lambda_{min}(Q+\lambda^{-1}PBB^TP)||x||^2 + 2x^TPB\sum_{i=0}^{m}\phi_i\tilde{w}_i\\
&+ 2x^TPB[\epsilon(e)+d(t)] + 2\sum_{i=0}^{m}(x^TPB\phi_i(e) + \frac{r_2}{r_1}\hat{w}_i)\tilde{w}_i,\\
&= -\lambda_{min}(Q+\lambda^{-1}PBB^TP)||x||^2 + 2x^TPB\epsilon(e) + 2\frac{r_2}{r_1}\sum_{i=0}^{m}\hat{w}_i\tilde{w}_i,\\
&= -\lambda_{min}(Q+\lambda^{-1}PBB^TP)||x||^2 + 2x^TPB[\epsilon(e)+d(t)]\\
&- 2\frac{r_2}{r_1}\sum_{i=0}^{m}\tilde{w}_i^2 + 2\frac{r_2}{r_1}\sum_{i=0}^{m}w_i\tilde{w}_i. \qquad (3.89)
\end{aligned}$$

Notice that

$$\begin{aligned}
2x^TPB[\epsilon(e)+d(t)] &\le \eta x^TPBB^TPx + \frac{1}{\eta}[\epsilon(e)+d(t)]^2,\\
&\le \eta x^TPBB^TPx + \frac{1}{\eta}[\epsilon_M + d_M]^2, \qquad (3.90)\\
2\frac{r_2}{r_1}w_i\tilde{w}_i &\le \frac{r_2}{\beta r_1}w_i^2 + \beta\frac{r_2}{r_1}\tilde{w}_i^2. \qquad (3.91)
\end{aligned}$$

Then, the time derivative of (3.89) can be written as:

$$\dot{V} \leq - [\lambda_{min}(Q + \lambda^{-1}PBB^TP) - \eta\lambda_{max}(PBB^TP)]||x||^2$$
$$- 2\frac{r_2}{r_1}\sum_{i=0}^{m}(1 - \frac{1}{2}\beta)\tilde{w}_i^2 + \frac{1}{\eta}[\epsilon_M + d_M]^2 + \frac{r_2}{\beta r_1}w_i^2$$

Let $\theta = [x^T\tilde{w}_0\tilde{w}_1...\tilde{w}_m]^T$ and it follows that:

$$\dot{V} \leq -2\gamma||\theta||^2 + \lambda_1, \quad (3.92)$$

where

$$\gamma = \frac{1}{2}min\{\lambda_{min}(Q + \lambda^{-1}PBB^TP)\eta\lambda_{max}(PBB^TP),$$
$$2\ \frac{r_2}{r_1}(1 - \frac{1}{2}\beta)\}, \quad (3.93)$$
$$\lambda_1 = \frac{1}{\eta}[\epsilon_M + d_M]^2 + \frac{r_2}{\beta r_1}w_i^2. \quad (3.94)$$

Clearly, $\gamma > 0$ for sufficiently small η, β. The following condition is obtained for $\dot{v}$ of v to be negative:

$$||\theta|| > (\frac{\lambda_1}{2\gamma})^{1/2}. \quad (3.95)$$

In order to show the boundedness of the states and weights, note from

$$\mu(||\theta||^2) \leq V \leq \nu(||\theta||^2), \quad (3.96)$$

where $\mu = min\{\lambda_{min}(P), 1\}, \nu = max\{\lambda_{max}(P), 1\}$, that

$$||\theta|| \leq [\frac{\nu}{\theta}||\theta(0)||^2e^{-2\gamma t/\nu} + \frac{\lambda_1\nu}{2\mu\gamma}(1 - e^{-2\gamma t/\nu})]^{-1/2}. \quad (3.97)$$

From (3.97), it may be concluded that the state and weights are bounded. This completes the proof.

Remark 3.4:

Although many literatures have discussed the control of the following non-linear system (Ge (1996) and Fabri and Kadirkamanathan (1996))

$$\dot{x}(t) = Ax(t) + Bu(t) + f(x), \quad (3.98)$$

or

$$\dot{x}(t) = f(x(t)) + g(x(t))u(t), \tag{3.99}$$

using the RBF concepts, their results cannot yet be generally applied to the process (3.69) in which the uncertainty is the form of $f(x, u)$. Another novelty addressed here is the marriage of the widely accepted PID control and nonlinear adaptive control. From a practical viewpoint, this is useful, since to the process operator, he only has to tune the PID control. The RBF adaptation is oblivious to him and it may be switched off if performance with PI alone is adequate.

3.8.4 Passivity of dynamical systems

From (3.72) the closed-loop error system is

$$\dot{e}(t) = -ae(t) - bK_i \int_0^t e(\tau)d\tau - bK_p e(t) - b\sum_0^m \tilde{w}_i\phi_i - b\varepsilon - bd(t), \tag{3.100}$$

as shown in Fig. 3.16.

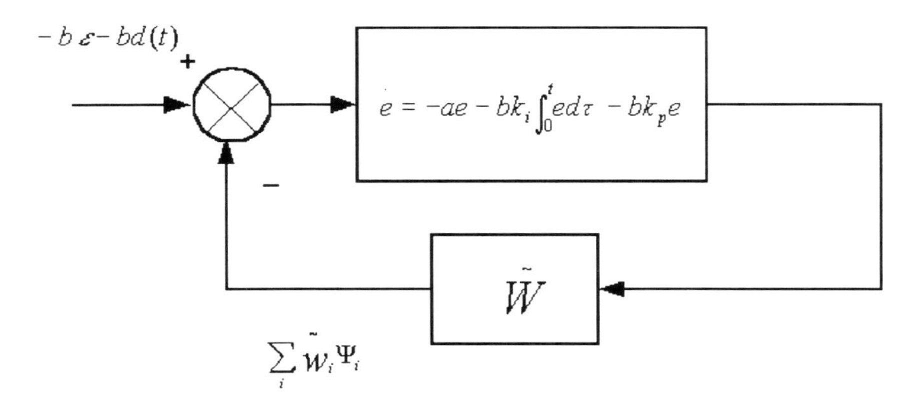

Fig. 3.16. Adaptive RBF error system

Passivity is essential in a closed-loop system as it guarantees the boundedness of signals, and hence adequate performance, even in the presence of additional unforseen disturbances as long as they are bounded.

A system with input $u(t)$ and output $y(t)$ is said to be passive (or to possess a passive mapping between u and y) if it verifies an equality of the so-called "power form" (Slotine and Li,1991).

$$\dot{V}(t) = y^T u - g(t), \tag{3.101}$$

with $V(t)$ lower bounded and $g(t) \geq 0$. That is:

$$\int_0^T y^T(\tau)u(\tau)d\tau \geq \int_0^T g(\tau) - \gamma_0^2, \tag{3.102}$$

for all $T \geq 0$ and some $\gamma_0 \geq 0$.

The system is said to be dissipative if it is passive and in addition:

$$\int_0^\infty y^T(\tau)u(\tau)d\tau \neq 0, \tag{3.103}$$

which implies

$$\int_0^\infty g(\tau)d\tau > 0.$$

A special form of dissipativity occurs if $g(t)$ is a monic quadratic function of $||x||$ with bounded coefficients, where $x(t)$ is the internal state of the system. Lewis et al. (1996) refer to this as State Strict Passivity (SSP).

It will be shown that the parameter tuning algorithm (3.85) yields desirable passivity properties of the RBF and the closed-loop system. Define

$$P = \begin{bmatrix} p_{11} & p_{12} \\ p_{21} & p_{22} \end{bmatrix}, \tag{3.104}$$

which is given by (3.80).

Theorem 3.6:

With the RBF parameter tuning algorithm (3.85), the mapping from $2p_{22}be(t)+2p_{12}b\int_0^t e(\tau)d\tau$ to $\sum_{i=0}^m \phi_i \tilde{w}_i$ is a SSP mapping.

Proof:

Defining a Lyapunov function:

$$V(t) = \frac{1}{r_1}\sum_{i=0}^m \tilde{w}^2, \tag{3.105}$$

and evaluating $\dot{V}$ yields

$$\dot{V} = \frac{2}{r_1}\sum_{i=0}^{m}\dot{\tilde{w}}_i\tilde{w}_i = 2\sum_{i=0}^{m}(-x^T PB\phi_i + \frac{r_2}{r_1}\hat{w}_i)\tilde{w}. \quad (3.106)$$

Since

$$\sum_{i=0}^{m}\hat{w}_i\tilde{w}_i = \sum_{i=0}^{m}(w_i - \tilde{w}_i)\tilde{w}_i \quad (3.107)$$

$$= \sum_{i=0}^{m}(w_i\tilde{w}_i - |w_i|^2) \leq \sum_{i=0}^{m}(w_M|\tilde{w}_i| - |\tilde{w}_i|^2), \quad (3.108)$$

and $x = [\int_0^t e(\tau)d\tau \;\; e(t)]^T$, it follows that:

$$\begin{aligned}
\dot{V} &\leq 2\sum_{i=0}^{m}\{\left[-\int_0^t e(\tau)d\tau \;\; -e(t)\right]\begin{bmatrix} p_{11} & p_{12} \\ p_{21} & p_{22}\end{bmatrix}\begin{bmatrix} 0 \\ -b\end{bmatrix}\phi_i\tilde{w}_i \\
&+ \frac{r_2}{r_1}(w_M|\tilde{w}_i| - |\tilde{w}_i|^2)\}, \\
&= \sum_{i=0}^{m}\{[2p_{22}be(t) + 2p_{12}b\int_0^t e(\tau)d\tau]\phi_i\tilde{w}_i - \frac{r_2}{r_1}(|\tilde{w}_i|^2 - w_M|\tilde{w}_i)\}, \\
&= \left[2p_{22}be(t) + 2p_{12}b\int_0^t e(\tau)d\tau\right]\sum_{i=0}^{m}\phi_i\tilde{w}_i - \frac{r_2}{r_1}(\sum_{i=0}^{m}|\tilde{w}_i|^2 \\
&- w_M\sum_{i=0}^{m}|\tilde{w}_i|),
\end{aligned}$$

which is in the power form with the last function quadratic in $\sum_{i=0}^{m}|\tilde{w}_i|$.

Thus, the parameter error block is SSP and the closed-loop error system is SSP. By using the passivity theorem, one may now conclude that the input/output signals of each block are bounded as long as the external inputs are bounded. Since the external inputs are bounded, we can conclude that the closed-loop signals are bounded.

3.8.5 Simulation study

In this section, two prominent industrial systems with nonlinear dynamics are simulated with the composite PI-adaptive control scheme to illustrate its effectiveness.

Nonlinear system I: Injection molding process

Injection molding is one of the major processes used in the plastics manufacturing industry. Control of the injection molding process involves three phases: filling, packing and holding, and cooling. For precise and consistent part production, it is very important to be able to accurately control the key injection molding machine variables during each phase. This application example will be mainly concerned with the process control of the filling phase. In Agrawal,Pandelidis and Pecht (1987), it has been shown that good control of the ram velocity is most essential during the filling phase. The entire phase may be regulated by controlling the injection speed of the ram so as to follow a pre-generated trajectory.

The reader is referred to Chiu (1991) and Rafizadeh et al. (1996) for the general theory of injection molding, and to Rafizaden et al. (1996) for the related control problems. A nonlinear dynamic model of the injection molding process governing the ram velocity to the hydraulic oil flow during the filling phase is described by:

$$\dot{z} = v_z, \tag{3.109}$$

$$\dot{P}_1 = \frac{\beta_1}{v_{10} + A_1 z}(u - A_1 v_z), \tag{3.110}$$

$$\dot{v}_z = \frac{1}{M}\left[P_1 A_1 - P_2 A_2 - 2\pi\eta R_n^{1-n}(l_0 + z)\left(\frac{(s-1)v_z}{k_r^{1-s} - 1}\right)^n\right], \tag{3.111}$$

$$\dot{P}_2 = \frac{\beta_2}{v_{20} - A_2 z}(A_2 v_z - Q_p), \tag{3.112}$$

$$y = z, \tag{3.113}$$

where z is the ram displacement, v_z is the ram velocity, P_1 is the hydraulic prssure, P_2 is the nozzle pressure, Q_p is the polymer flow rate, and u is the hydraulic oil flow into the injection cylinder. β_1, β_2, v_{10}, v_{20}, A_1, A_2, M, η, R_n, l_0, s, k_r, and n are the system parameters. The physical interpretation of these parameters and their values are shown in Tan et al.(1999). It is assumed that these parameters are unknown and to be determined.

Following Tan et al. (1999), it is first assumed that during the filling phase, the polymer melt flow Q_p is assumed to be constant. Relay tuning is first used to obtain a first-order model as given by:

$$\tilde{G}_p(s) = \frac{2.96 \times 10^{-4}}{1 + 0.0165s}. \tag{3.114}$$

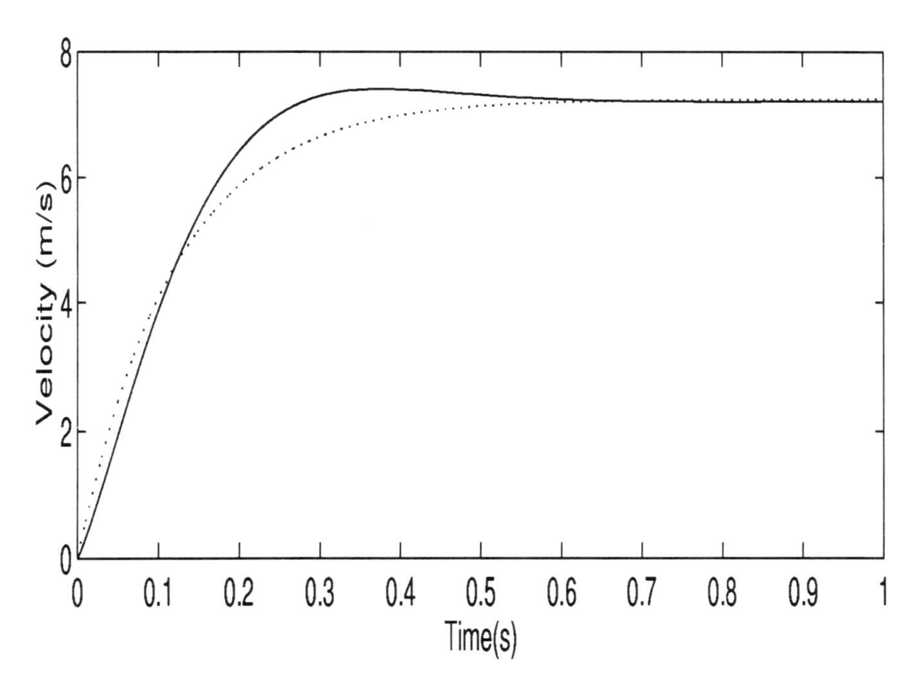

Fig. 3.17. Comparison between linear model(dashed line) and actual nonlinear system(solid line)

Fig. 3.17 shows the response of (3.114) which is quite close to that of the nonlinear system. Based on this model, the PI parameters are designed using (3.79). The following parameters are used in the design:

$$Q = 10^7 \begin{bmatrix} 10^4 & 0 \\ 0 & 1 \end{bmatrix}, \tag{3.115}$$

$$\lambda = 1. \tag{3.116}$$

PI control, designed with respect to the linear model, is given by:

$$u_{pi}(t) = 4150e(t) + 316230 \int_0^t e(\tau)d\tau \tag{3.117}$$

From Theorem 3.5, the adaptive part of the controller is obtained, where m is chosen as $m = 9$, the center c_i of the RBF basis are $[0.01\ 10], [-0.01\ -10], [0.03\ 20], [-0.03\ -20], [0.06\ 30], [-0.06\ -30], [0.09\ 40], [-0.09\ -40], [0.12\ 50], [-0.12\ -50]$ and $\sigma = 500$. Fig. 3.18 shows the simulation results. If only the pure PI control is used in the controller, the results are shown in Fig. 3.19. In contrast, the enhancement from RBF adaptation is not too obvious as the dominant model is already a very close approximation to the nonlinear system. In this case, the adaptive component may be disabled.

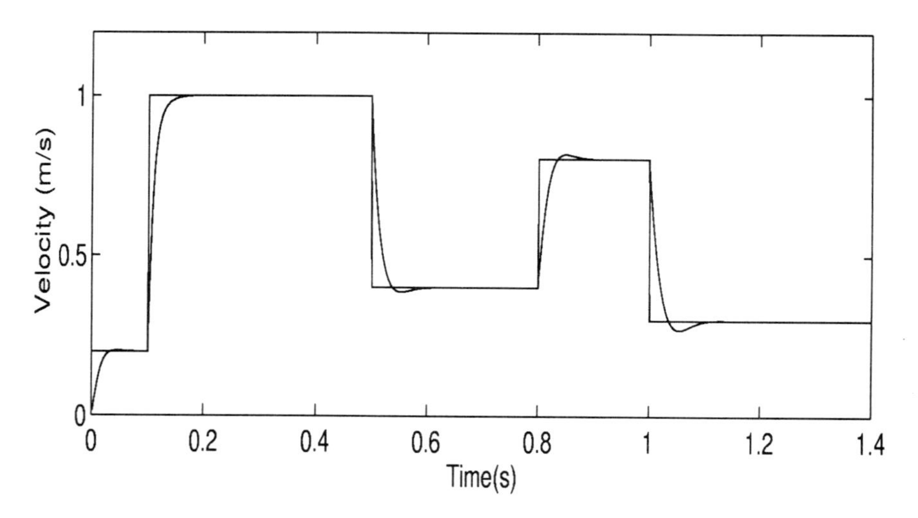

Fig. 3.18. Response with the composite PI-adaptive controller

However, in a practical injection molding process, Q_p is not a constant as assumed earlier (Rafizadeh et al., 1997). It is proportional to the cavity pressure whose response during the entire process is shown in Fig. 3.20. Under this new condition, the dominant model is much less adequate (this case is modeled in the simulation by letting $Q_p = 1000exp(4t)$). Fig. 3.21 shows that the fixed PI control fails to achieve good tracking performance. In contrast, the PI-adaptive controller still recovers the good performance as shown in Fig. 3.22.

For time-varying setpoints, the PI-adaptive method can still be applied to designing the controller although it is difficult to prove this claim. It will instead be illustrated via an example.

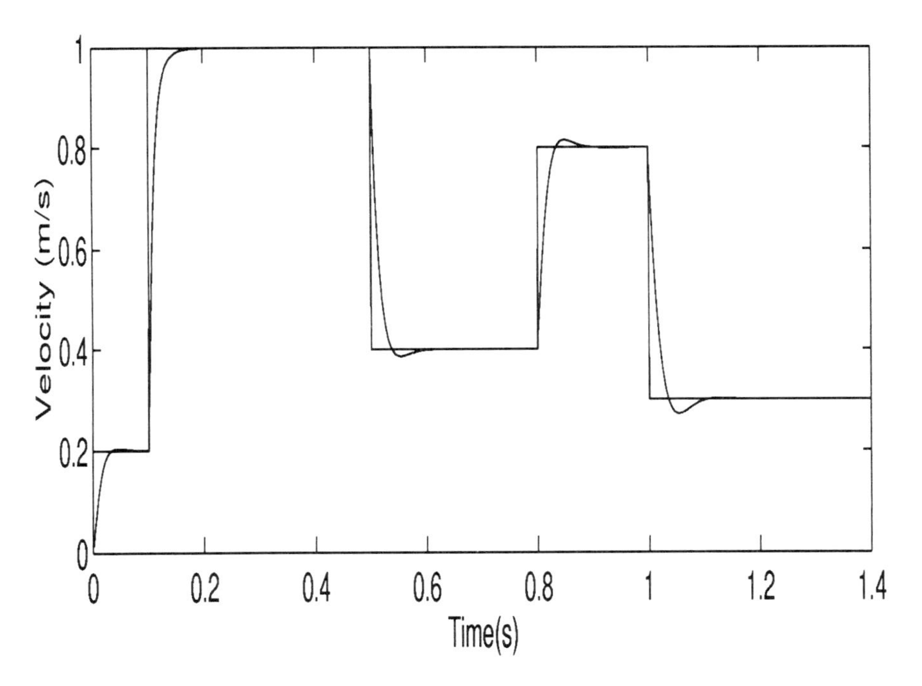

Fig. 3.19. Response with pure PI control

Nonlinear system II: Linear motor

Linear motors are becoming very popular for applications requiring high speed, high accuracy operations due to their mechanical simplicity. For these kinds of applications, the control requirements are particularly stringent which makes the control strategy formulation very challenging.

The linear motor considered here is a brushed permanent magnet DC linear motor produced by Anorad Corp. A nonlinear model can be used to describe the dynamics of the linear motor:

$$\dot{x} = v, \tag{3.118}$$

$$\dot{v} = \frac{u - f_{friction} - f_{ripple}}{m} + w(t), \tag{3.119}$$

where $f_{friction}$ is the friction force, f_{ripple} is the ripple force, u is the developed force, m is the combined mass of translator and load, and w represents any other residual disturbances.

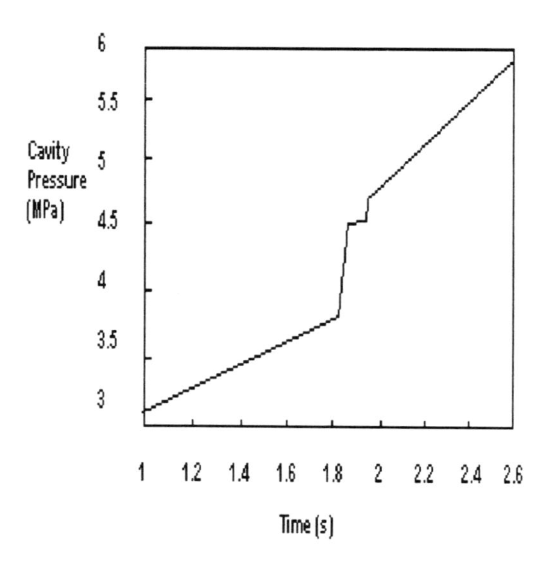

Fig. 3.20. Profile of cavity pressure during a molding cycle

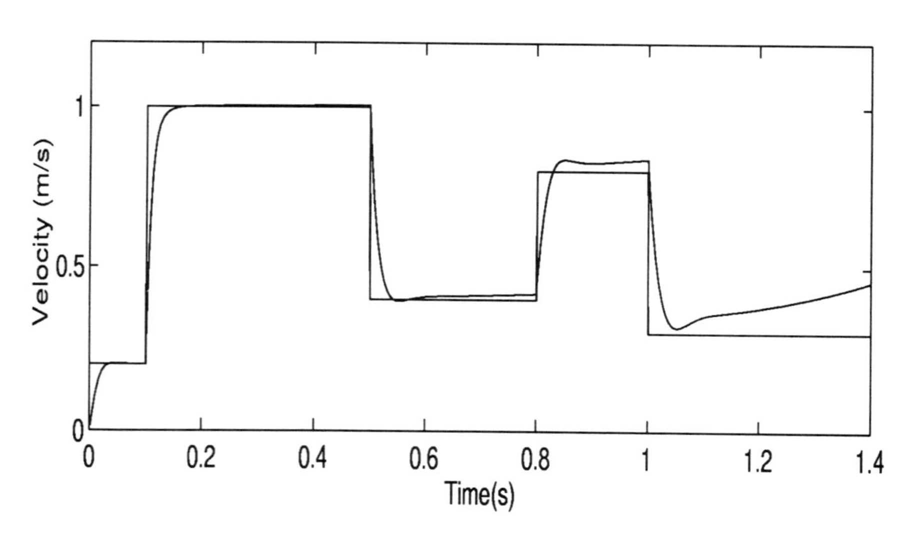

Fig. 3.21. Performance with fixed PI when $Q_p = 1000e^{4t}$

The friction and ripple forces are assumed to be modeled as follows:

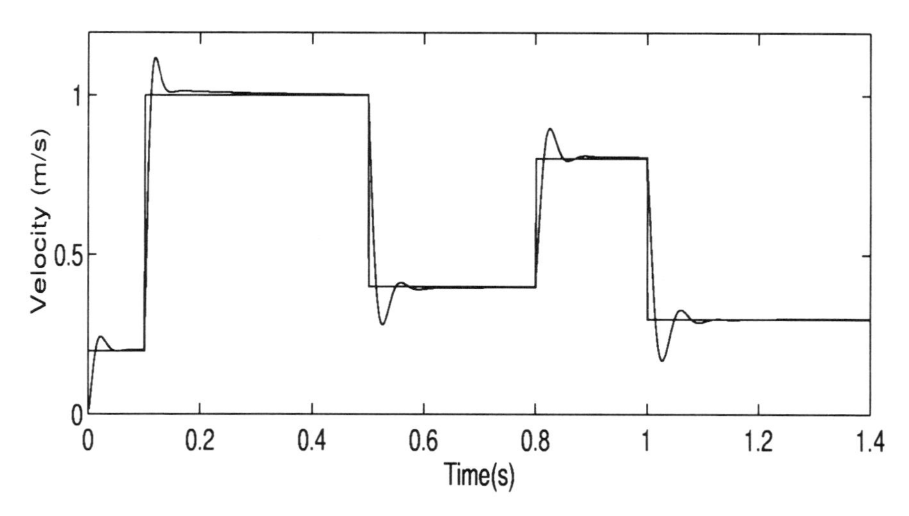

Fig. 3.22. Performance with the PI-adaptive controller when $Q_p = 1000e^{4t}$

$$f_{friction} = (f_c + (f_s - f_c)e^{-(\dot{x}/\dot{x}_s)^\delta} + f_v\dot{x})sgn(\dot{x}), \tag{3.120}$$

$$f_{ripple} = b_1 sin(w_0 x). \tag{3.121}$$

The above model allows the evaluation of the friction force during both sticking and slipping motions. The model parameters can be found in Tan et al. (1998a).

The desired trajectory is planned as follows:

$$v_d(\tau) = (x_0 - x_f)(60\tau^3 - 30\tau^4 - 30\tau^2), \tag{3.122}$$

where $\tau = t/(t_f - t_0)$, x_0 is the initial position and x_f is the final position. In this simulation study, the parameters are chosen as $x_0 = 0, x_f = 1.0m, t_f = 1s$.

To apply the PI-adaptive controller, the system (from control voltage to velocity) is first modeled as:

$$\tilde{G}_p(s) = \frac{0.6014}{s + 8.3} \tag{3.123}$$

This model is not ideal as shown in the response of Fig. 3.23 due to the nonlinear effects obviously neglected in the linear model. The PI control law is designed according to Theorem 3.5 as:

$$u_{pi}(t) = 442.6629e(t) + 83 \int_0^t e(\tau)d\tau. \tag{3.124}$$

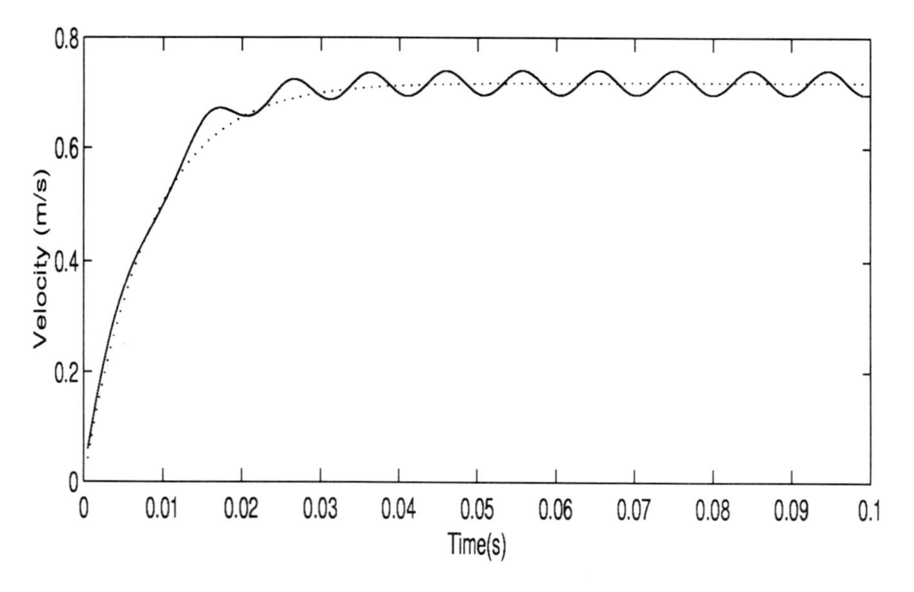

Fig. 3.23. Comparison between linear model (dashed line) and the actual nonlinear system (solid line)

The RBF structure is the same as that used in the injection molding simulation. Fig. 3.24 shows the control results. The simulation results for the pure PI control are shown in Fig. 3.25.

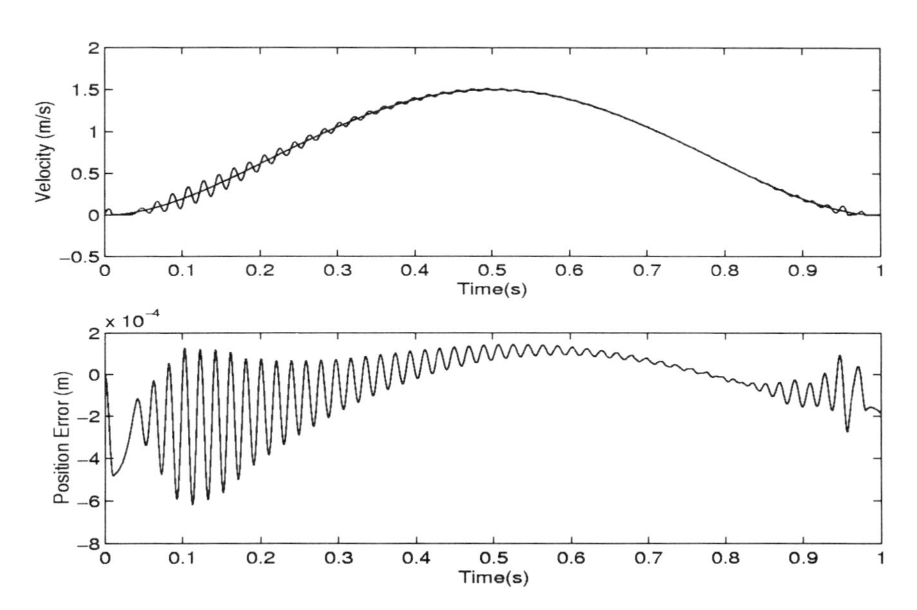

Fig. 3.24. Tracking performance with the PI-adaptive control for the linear motor

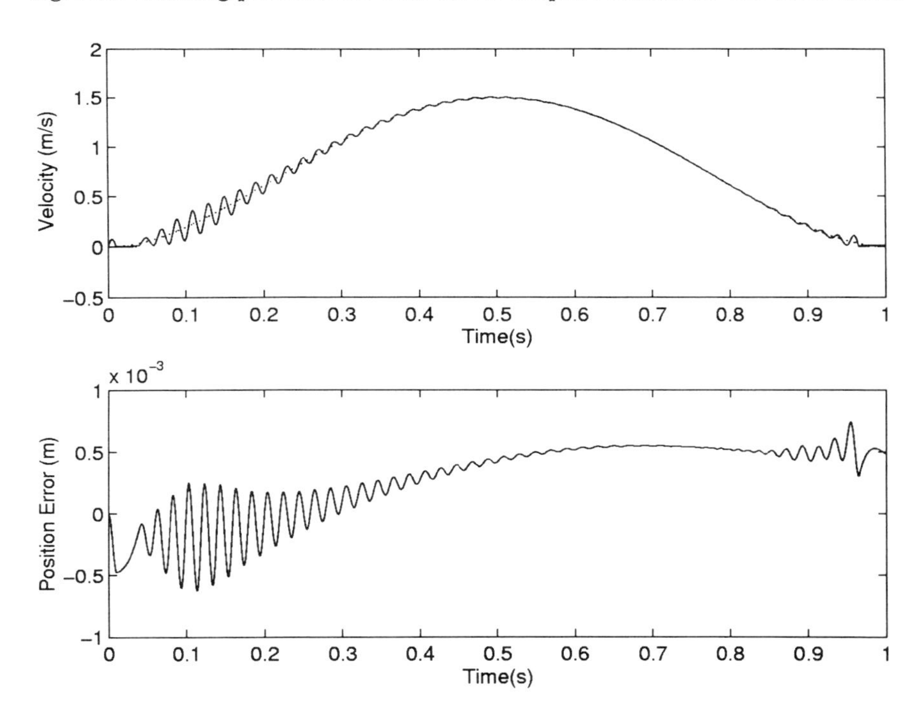

Fig. 3.25. Tracking performance with PI control for the linear motor

CHAPTER 4
AUTOMATIC TUNING

4.1 Introduction

Manufacture of PID controllers has undergone many changes over the last decades. PID controllers have evolved from a pneumatic setup in the 1940's to electrical ones and now modern microprocessors based devices. While the implementation has changed tremendously, the basic functions of controllers, the PID algorithm, has not and probably will never changed. For the user, the job of tuning the controller has also been essentially the same throughout the decades, although new PID designs have been continuously developed to cater to different sets of control requirements.

However, new possibilities and functionalities have become possible with a microprocessor-driven PID controller. Process controllers today often contain much more than just the basic PID algorithm. Fault diagnosis, alarm handling, signals scaling, choice of type of output signal, filtering, simple logical and arithmetic operations are becoming common functions to be expected in modern PID controllers. Physical size of the controller has shrunk compared to the analog predecessors and yet the functions and performance have greatly increased. Furthermore, riding on the advances in adaptive control and techniques, the modern PID controllers are becoming intelligent. Many high-end controllers are appearing in the market equipped with auto-tuning and self-tuning features. No longer is tedious manual tuning an inevitable part of process control. The role of operators in PID tuning has been very much reduced to simple specifications and decision.

In Chapter 2 and 3, different systematic methods for tuning PID controllers are described. Irregardless of the design method, the following three phases are applicable:

1. The process is disturbed with specific control inputs or control inputs automatically generated in the closed-loop.
2. The response to the disturbance is analyzed, yielding a model of the process which may be non-parametric or parametric.

3. Based on this model and certain operation specifications, the control parameters are determined.

Automatic tuning of controllers means quite simply that the above procedures are automated so that the disturbances, model calculation and choice of controller parameters all occur within the same controller. In this way, the work of the operator is made simpler, so that instead of having to calculate suitable controller parameters himself, he only needs to initiate the tuning process. He may have to give the controller some information about the process before the tuning is done, but this information will be considerably simpler to specify than the controller parameters.

Automatic tuning of PID controllers can be done in many different ways. The process can be disturbed in different ways, e.g. via steps, pulses or some form of frequency analysis. The model built up can also take many different forms. The principles behind two main types are described, one based on time-domain analysis of the step response of the process and the other based on frequency response analysis.

4.1.1 Methods based on step response analysis

Most methods for automatic tuning of controllers are based on disturbance of the process by a step change in the control signal. When the operator wishes to tune his controller, the control signal is changed in steps, and the controller records how the process variable reacts to the disturbance. When the process variable has reached its new equilibrium level, a model is calculated. When the model has been calculated, the controller parameters are chosen. This is usually done by variations of the Ziegler-Nichols and other methods as presented in Chapter 2, i.e., via a simple look-up table where the controller gain is chosen in inverse proportion to the gain of the process and the integral and derivative times are chosen in proportion to the dead-time and dominant time constant of the process.

The greatest difficulty in carrying out the tuning process automatically is in selecting the size of the step disturbance. The user will naturally want the disturbance to be as small as possible so that the process is not disturbed more than necessary. The controller, on the other hand, finds it easier to determine the process model if the disturbance is large. The disturbance must under all circumstances be large enough so that the step response of the process variable can be distinguished from the prevailing noise.

The noise level can be calculated by observing the variations in the process variable. Even if the noise level is known, it is still difficult to specify the step size of the control signal unless the static gain of the process is known. The result of this dilemma is usually that the operator himself has to decide how large the step in the control signal should be. In spite of this, instead of the operator having to tune the controller parameters, he only needs to specify the acceptable change in control signal.

Controllers with automatic tuning which are based on this technique have become very common in the last few years.

4.1.2 Methods based on frequency response analysis

Automatic tuning of PID controllers can also be carried out where the process dynamics are determined by the use of frequency response analysis instead of step response analysis. The greatest problem in step response analysis has been to determine the size of the step in the control signal. With frequency analysis, the corresponding problem is to determine the amplitude of the oscillation in the control signal. However, the problem is not as great as with step response analysis, since the amplitude of the oscillations may be adjusted during the course of the experiment. For example, if the amplitude of the process variable is observed to be too large, the amplitude of the control signal can be reduced without having to terminate the experiment.

The main problem with frequency analysis is probably in determining a suitable frequency, i.e., a frequency where a good knowledge of the process is essential. This frequency range is the area around the ultimate frequency, i.e., the range where the process has a phase shift of $-\pi$.

The Ziegler-Nichols frequency response method involves a method of automatically generating a frequency range of interest. Unfortunately, this method is not easy to automate, as it means operating to the stability limit. Also, in most cases, the method results in the amplitude of the process variable being too great.

There is another method for finding a suitable frequency for the frequency analysis automatically, and for generating this oscillation. This is the relay feedback method which has become a common and popular way for automatic tuning of controllers. Full sections will be devoted to the relay feedback method in this chapter.

4.2 Step Response Approach

Several typical identification methods based on step tests are used in industry (Shinskey, 1988; Astrom and Hagglund, 1995; Marlin, 1995). Most of them result in the first-order plus dead-time model:

$$G_p(s) = \frac{K_p}{Ts+1}e^{-Ls}, \tag{4.1}$$

which describes a linear monotonic process quite well in most chemical processes and HVAC processes, and is often sufficient for PID controller tuning (Hang and Chin, 1991; Halevi, 1991).

For the identification of model (4.1), Astrom and Hagglund (1995) proposed a graphical method. The process static gain K_p is obtained from the steady states of the process input and output. The intercept of the tangent to the step response that has the largest slope with respect to the horizontal axis gives L. T is determined from the difference between L and the time when the step response reaches the value of $0.63K_p$. In a two-point method (Marlin, 1995), the static gain K_p is obtained as before. The times at which the process output reaches 28% of K_p and 63% of K_p respectively are measured, and used to determine T and L.

These two methods are simple, but they are quite sensitive to large measurement noise. The area-based methods may have better estimation robustness. Suppose that the static gain K_p is determined as before, and the average residence time T_{ar} is computed from the area A_0 in Fig. 4.1 as:

$$T_{ar} = \frac{A_0}{K_p} = \frac{\int_0^\infty [y(\infty) - y(t)]dt}{K_p}.$$

The area A_1 under the step response up to time T_{ar} is also measured. Then T and L can be estimated as:

$$T = \frac{eA_1}{K_p} = \frac{e\int_0^{T_{ar}} y(t)dt}{K_p},$$

$$L = T_{ar} - T = \frac{A_0}{K_p} - \frac{eA_1}{K_p}.$$

This method is less sensitive to high-frequency noise (Astrom and Hagglund, 1995) compared to the two step methods above, with which the model is determined from only a few values of the step response. However, the estimation accuracy depends mainly on the A_0. In order to have an accurate A_0,

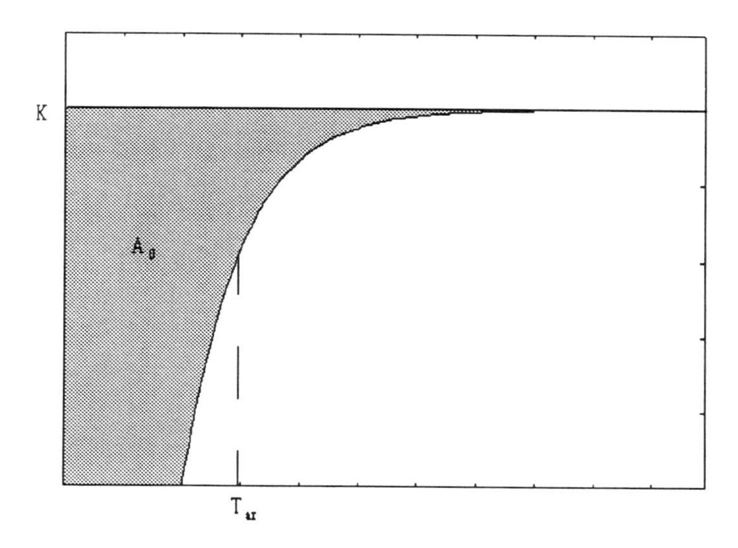

Fig. 4.1. Area method for a monotonic step response

one needs a test duration that is long enough for the process to enter the new steady state completely. Some other step-estimation methods based on calculating the areas of the response curve have also been reported (Rake, 1980). However, these methods are only viable for processes without dead-time.

A method based on step tests is presented in this section which can overcome the limitations in the above identification methods. The parameters of a first-order plus dead-time continuous model are obtained directly from a set of newly derived linear regression equations. No iteration is needed, and the method is robust to measurement noise. The method does not need to wait for the process to enter completely into the new steady states, thus reducing the testing time. The method works well for industrial processes that are sufficiently described by a first-order plus dead-time model. Simulation and a real-time implementation will show the effectiveness of the method and estimation improvement over the existing methods.

4.2.1 Modeling from a step test

Suppose a given process is in zero initial state, before a step change with amplitude of h at $t = 0$ in the process input $u(t)$ is applied. The process input $u(t)$ and the resulting output response $y(t)$ are logged until the process enters the new steady state. For a process described by (4.1), the output transient $y(t)$ after $t = L$ is described by:

$$y(t) = hK_p(1 - e^{-\frac{t-L}{T}}) + \omega(t), \qquad \text{for } t \geq L, \tag{4.2}$$

where $\omega(t)$ is the white noise present in the measurement of $y(t)$. (4.2) is re-written as:

$$e^{-\frac{t-L}{T}} = 1 - \frac{y(t)}{hK_p} + \frac{\omega(t)}{hK_p}, \qquad \text{for } t \geq L. \tag{4.3}$$

Integrating $y(t)$ in (4.2) from $t = 0$ to $t = T (T \geq L)$ yields:

$$\int_0^T y(t)dt = hK_p(t + Te^{-\frac{t-L}{T}})\Big|_L^T + \int_0^T \omega(t)dt.$$

Using (4.3) and $y(L)$=0 leads to

$$\int_0^T y(t)dt = hK_p[T - L - T\frac{y(T)}{hK_p}] + [T\omega(t)]\Big|_L^T + \int_0^T \omega(t)dt. \tag{4.4}$$

Let

$$A(T) = \int_0^T y(t)dt, \quad \delta(T) = [T\omega(t)]\Big|_L^T + \int_0^T \omega(t)dt.$$

Then (4.4) can be re-written as:

$$A(T) = hK_p[T - L - T\frac{y(T)}{hK_p}] + \delta(T),$$

or

$$\begin{bmatrix} hT & -h & -y(T) \end{bmatrix} \begin{bmatrix} K_p \\ LK_p \\ T \end{bmatrix} = A(T) - \delta(T). \tag{4.5}$$

By collecting (4.5) for all sampled $y(T)$ after $T \geq L$, a system of linear equations is obtained as:

$$\Psi\Theta = \Gamma + \Delta, \qquad \text{for } \ T \geq L, \tag{4.6}$$

where $\Theta = [K_p \; LK_p \; T]^T$,

$$\Psi = \begin{bmatrix} hmT_s & -h & -y[mT_s] \\ h(m+1)T_s & -h & -y[m+1]T_s \\ \vdots & \vdots & \vdots \\ h(m+n)T_s & -h & -y[m+n]T_s \end{bmatrix}, \Gamma = \begin{bmatrix} A[mT_s] \\ A[(m+1)T_s] \\ \vdots \\ A[(m+n)T_s] \end{bmatrix},$$

$$\Delta = \begin{bmatrix} -\delta[mT_s] \\ -\delta[(m+1)T_s] \\ \vdots \\ -\delta[(m+n)T_s] \end{bmatrix}.$$

T_s is the sampling interval, and $mT_s \geq L$. The estimation $\tilde{\Theta}$ of Θ in (4.6) can be obtained using the least-squares method as:

$$\tilde{\Theta} = (\Psi^T \Psi)^{-1} \Psi^T \Gamma. \tag{4.7}$$

K_p, T and L in (4.1) can then be calculated from $\tilde{\Theta}$.

In the noise-free testing environment, the above estimator yields true parameters. However, in the presence of a large amount of noise, (4.7) is biased (Young, 1970; Strejc, 1980), because $\delta(T)$ is a zero-mean correlated noise instead of a white one. One solution is to use the instrumental variable (IV) least-squares method (Young, 1970; Strejc, 1980). The instrumental matrix Z is selected such that:

(a) The inverse of

$$\lim_{n \to \infty} \frac{1}{n} Z^T \Psi,$$

exists, and

(b)

$$\lim_{n \to \infty} \frac{1}{n} Z^T \Delta = R_{Z\Delta} = 0,$$

i.e., Z and Δ are uncorrelated.

Many solutions for Z exist. In this case, Z is chosen as:

$$Z = \begin{bmatrix} mT_s & -1 & \frac{1}{mT_s} \\ (m+1)T_s & -1 & \frac{1}{(m+1)T_s} \\ \vdots & \vdots & \vdots \\ (m+n)T_s & -1 & \frac{1}{(m+n)T_s} \end{bmatrix}.$$

With this instrumental matrix, the best estimation $\tilde{\Theta}$ of Θ in (4.6) is obtained as:

$$\tilde{\Theta} = (Z^T \Psi)^{-1} Z^T \Gamma. \tag{4.8}$$

Remark 4.1:

It is noted that (4.5) is effective only after T=L. In practice, the selection of the logged $y(T)$ after T=L can be made as follows. When the process enters a zero initial state, the process output will be monitored for a period (the 'listening period'), during which the noise band B_n can be determined. Then, the step test starts until the process settles down again. $y(T)$ satisfying

$$abs(y(T)) > 2B_n,$$

can be treated as the process response after T=L, and thus be used in (4.8). Another detector is to check if $y(T)$ satisfies

$$abs(y(T)) > 0.1 * abs(y[(m+n)T_s]).$$

Remark 4.2:

The method (henceforth referred to as IV-LS (instrumented variable-least squares)) obtains the process continuous transfer function model directly from the process response under a step test, so no iteration is needed. The continuous transfer function model is familiar to most control engineers, and is a sufficient basis on which to adopt many existing tuning rules to design a PID/PI controller.

Remark 4.3:

The response, obtained by stimulating the estimated model using the same step signal as in the test, can be visually or statistically compared with the response from the test, and used to validate the estimated model.

Remark 4.4:

This modeling method from a step test can be easily modified to give a recursive version. The recursive version of an instrumental variable least-squares method has been reported (Young, 1970; Strejc, 1980).

4.2.2 Simulation study

The step identification method is now applied to several typical processes. For a better assessment of its accuracy, identification errors in both the time domain and the frequency domain are considered. This is because some step identification methods are found to be able to fit the process response well in the time domain, but the frequency response of the model sometimes deviates too far from the real process frequency response. To achieve better control performance, the estimation error should be small in both time and frequency domains. The comparison is made with the area method (Astrom and Hagglund 1995), to show the performance enhancement.

The time-domain identification error is measured by the standard deviation:

$$\varepsilon = \frac{1}{n+1} \sum_{k=m}^{m+n} [y(kT_s) - \tilde{y}(kT_s)]^2, \tag{4.9}$$

where $y(kT_s)$ is the actual process output under a step change, while $\tilde{y}(kT_s)$ is the response of the estimated process under the same step change. Without loss of generality, a unit step is employed in the following simulation.

The frequency-domain identification error is measured by the *worst-case error*:

$$E = \max_i \left\{ \left| \frac{\tilde{G}_p(j\omega_i) - G_p(j\omega_i)}{G_p(j\omega_i)} \right| \times 100\%, \; i = 1, 2, \cdots, \mathrm{M} \right\}, \tag{4.10}$$

where $G_p(j\omega_i)$ and $\tilde{G}_p(j\omega_i)$ are the actual and estimated process frequency responses respectively. The Nyquist curve for a phase ranging from 0 to π is considered, since this part is the most significant for control design.

Example 4.1:

Consider a first-order plus dead-time process:

$$G_p(s) = \frac{e^{-s}}{s+1}.$$

A unit step test is performed, and the process input and output in the step test are recorded. Recall that the process response does not have to fully enter the steady state before the method cabn be applied. Here, the response from $t = 0$ to $t = 4.14$ is used, with $y(t = 4.14) = 0.95$. Using (8), the estimated process model is obtained as:

$$\tilde{G}_p(s) = \frac{e^{-s}}{0.997s + 1},$$

which is almost identical to the actual process. The identification errors are ε=7.5×10^{-7} and E=0.35% (Table 4.1).

Table 4.1. Comparison of identification results

	The IV-LS Method			The Area Method		
Process	$\tilde{G}_p(s)$	ε	E	$\tilde{G}_p(s)$	ε	E
$\frac{e^{-s}}{s+1}$	$\frac{1.00e^{-1.00s}}{0.997s+1}$	7.5×10^{-7}	0.35%	$\frac{e^{-1.01s}}{0.99s+1}$	1.8×10^{-6}	2.4%
$\frac{e^{-4s}}{(10s+1)(2s+1)}$	$\frac{1.03e^{-5.47s}}{11.41s+1}$	4.5×10^{-5}	10.0%	$\frac{e^{-5.83s}}{10.19s+1}$	2.1×10^{-5}	15.8%
$\frac{1-s}{(s+1)^5}$	$\frac{1.01e^{-3.73s}}{2.45s+1}$	1.9×10^{-3}	4.8%	$\frac{e^{-4.00s}}{2.11s+1}$	9.0×10^{-4}	12.5%
$\frac{1}{(s+1)^8}$	$\frac{1.06e^{-4.94s}}{3.81s+1}$	7.7×10^{-4}	10.1%	$\frac{e^{-4.3s}}{4.3s+1}$	2.2×10^{-3}	28%

For the area identification method (Astrom and Hagglund, 1995), if the step response from $t = 0$ to the time when the process enters the new steady state well, with $y(8.28s)$=0.999, is used in computation, then the estimated model is

$$\tilde{G}_p(s) = \frac{e^{-1.01s}}{0.99s + 1}.$$

The estimation errors are $\varepsilon = 1.8\times10^{-6}$ and E=2.4% (Table 4.1). However, if the response in $t = 0 \sim 4.14s$ is used, then the estimation will deteriorate to

$$\tilde{G}_p(s) = \frac{0.95e^{-1.07s}}{0.78s + 1},$$

with $\varepsilon = 3.7\times10^{-4}$ and E=15.6%.

It is clear that the method saves a third to one half of the testing time, and gives a more accurate result. In the subsequent examples, the step responses entering the steady state are used in the area method.

Example 4.2:

Consider a second order plus dead-time process:

$$G_p(s) = \frac{e^{-4s}}{(10s+1)(2s+1)}.$$

The results from the IV-LS method and the area method are both shown in Table 4.1. An improvement is observed with the identification results from the IV-LS method.

Example 4.3:

The IV-LS method is also tested on a non-minimum-phase process:

$$G_p(s) = \frac{1-s}{(s+1)^5}.$$

The difference between the actual and the identified process is very small. The result from the area method is also listed for comparison.

Example 4.4:

Consider a process with multiple lag:

$$G_p(s) = \frac{1}{(s+1)^8},$$

which is used in Astrom and Hagglund (1995). The results from both methods are presented in Table 4.1. Clearly, the IV-LS method is more accurate.

Since the IV-LS method makes use of multiple points rather than one or two points on the process response, and adopts an instrumental variable least-squares technique, it is expected to be robust to noise. To demonstrate this, the process of $G_p(s) = \frac{e^{-s}}{s+1}$ in Example 4.1 is again tested in the noisy case. In the context of system identification, the noise-to-signal ratio is usually defined as:

$$\text{NSR} = \frac{mean\ power\ spectrum\ density\ of\ noise}{mean\ power\ density\ spectrum\ of\ signal}.$$

Under measurement noise with NSR=25%, a step test is performed. The process response is shown in Fig. 4.2. The model obtained by the IV-LS method is $\tilde{G}_p(s) = \frac{1.00e^{-1.01s}}{1.01s+1}$, which is almost the same as the actual process.

Table 4.2. Identification results under different noise levels

NSR	0	15%	25%	48%	65%
$\tilde{G}_p(s)$	$\frac{1.000e^{-1.000s}}{0.997s+1}$	$\frac{1.003e^{-1.005s}}{1.003s+1}$	$\frac{1.00e^{-1.01s}}{1.01s+1}$	$\frac{1.01e^{-1.00s}}{1.02s+1}$	$\frac{1.02e^{-1.00s}}{1.06s+1}$
E	0.35%	1.38%	1.91%	2.76%	3.87%

The step response of the estimated model is also plotted in Fig. 4.2, which shows that the estimation agrees very well with the actual one. The estimation error E is 1.91%. The estimation results of the process under different noise levels are listed in Table 4.2.

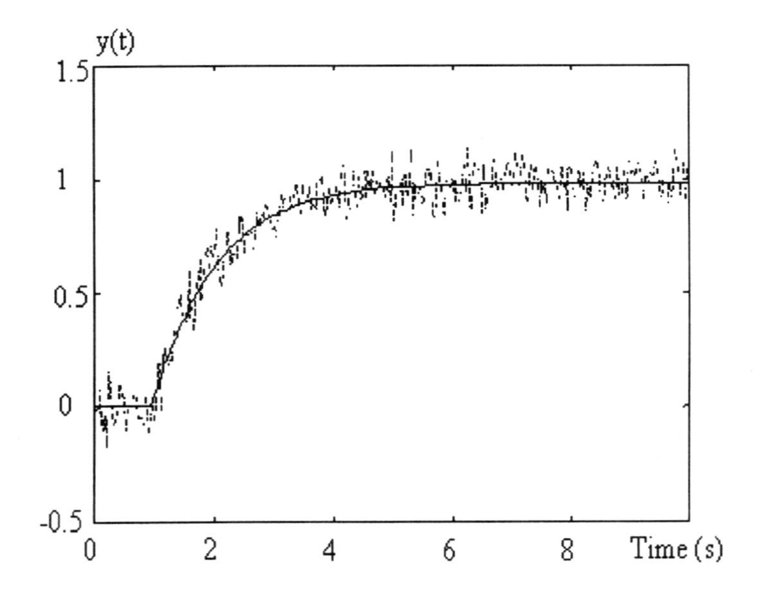

Fig. 4.2. The process and model responses under a step change
(— Actual process response under noise influence; - - - Estimated model response)

4.3 Relay Feedback Approach

An interesting experiment design for process frequency response analysis is the relay feedback system shown in Fig. 4.3, first pioneered by Astrom and co-workers.

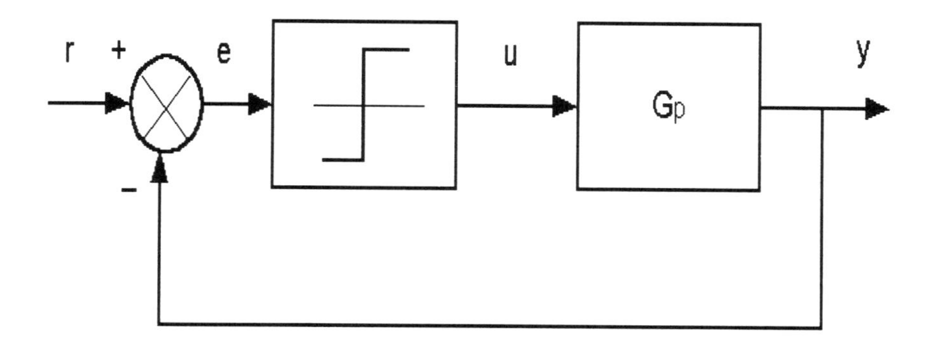

Fig. 4.3. Relay feedback system.

This method has been the subject of much interests in recent years and it has been field tested in a wide range of applications. There are many attractive features associated with the relay feedback technique. First, for most industrial processes, the arrangement automatically results in a sustained oscillation approximately at the ultimate frequency of the process. From the oscillation amplitude, the ultimate gain of the process can be estimated. This alleviates the task of input specification from the user and therefore is in delighting contrast to other frequency-domain based methods requiring the frequency characteristics of the input signal to be specified. This particular feature of the relay feedback technique greatly facilitates automatic tuning procedures, since the arrangement will automatically give an important point of the process frequency response. Secondly, the relay feedback method is a closed-loop test and the process variable is maintained around the set-point value. This keeps the process in the linear region where the frequency response is of interest and this works well on highly non-linear processes; the process is never pushed very far away from the steady-state conditions. Current identification techniques relying on transient analysis such as impulse or step tests do not possess this property, and they are therefore ineffective for processes with non-linear dynamics. Thirdly, the relay feedback technique does not require prior information of the system time constants for a careful choice of the sampling period. The choice of the sampling period has always been a tricky problem for traditional parameter estimation techniques.

If the sampling interval is too long, the dynamics of the process will not be adequately captured in the data and the accuracy of model subsequently obtained will be poor as a consequence. While a conservative safety-first approach towards this decision may be to select the smallest sampling period supported by the data acquisition equipment, this would result in too much data collection with inconsequential information. A corrective action then is data decimating in the post-treatment phase which for real-time parameter estimation may not be tolerable. Spared of these cumbersome and difficult decisions, the relay feedback method is therefore an attractive method to consider in auto-tuning applications.

Relay systems can be traced back to their classical configurations. In the fifties, relays were mainly used as amplifiers but such applications are obsolete now, owing to the development of electronic technology. In the sixties, relay feedback was applied to adaptive control. One prominent example of such applications is the self-oscillating adaptive controller developed by Minneapolis Honeywell which uses relay feedback to attain a desired amplitude margin. This system was tested extensively for flight control systems, and it has been used in several missiles. It was in the eighties that Astrom successfully applied the relay feedback method to auto-tune PID controllers for process control, and triggered a resurgence of interest in relay methods, including extensions of the method to more complex systems. A recent survey of relay methods is provided in Astrom (1995).

This section is focused on non-parametric frequency response identification using a relay feedback. A review of the basic relay method is first provided, followed by variants of the basic method which expand its applicability to more scenarios which may require better accuracy or faster tuning time. The basic relay feedback method is essentially an off-line tuning method with several associated difficulties. An online PID auto-tuning method is further illustrated using relay feedback without having to break the control loop. Within the same experiment, the PID control structure and existing control parameters can be identified, and the control performance assessed to determine if re-tuning is necessary. Next, a method for online estimation of frequency response using *Discrete Fourier Transform* techniques is illustrated. Frequency response estimation in the non-parametric form may not be directly useful, either for simulation or model-based control purposes. The subsequent section in the chapter elaborates how common low-order rational transfer function models may be easily fitted to the frequency response obtained from the relay-based identification methods. These models may then be used for modern, advanced and model-based PID designs. Finally, in the concluding section, methods for self-tuning and adaptation of the PID controller based on naturally occurring transients are presented.

4.3.1 Basic idea

The ultimate frequency ω_π of a process, where the phase lag is $-\pi$, can be determined automatically from an experiment with relay feedback as shown in Fig. 4.3.

The usual method employed to analyze such systems is the describing function method which replaces the relay with an "equivalent" linear time-invariant system. For estimation of the critical point (ultimate gain and ultimate frequency), the self-oscillation of the overall feedback system is of interest. Here, for the describing function analysis, a sinusoidal relay input,

$$e(t) = a \sin \omega t,$$

is considered, and the resulting signals in the overall system are analyzed. The relay output $u(t)$ in response to $e(t)$ would be a square wave having a frequency ω and an amplitude equal to the relay output level μ. Using a Fourier series expansion, the periodic output $u(t)$ can be written as

$$u(t) = \frac{4\mu}{\pi} \sum_{k=1}^{\infty} \frac{\sin(2k-1)\omega t}{2k-1}.$$

The describing function of the relay $N(a)$ is simply the complex ratio of the fundamental component of $u(t)$ to the input sinusoid, i.e.

$$N(a) = \frac{4\mu}{\pi a}.$$

Since the describing function analysis ignores harmonics beyond the fundamental component, define here the residual ϱ as the entire sinusoidally-forced relay output minus the fundamental component, i.e. the part of the output that is ignored in the describing function development,

$$\varrho = \frac{4\mu}{\pi} \sum_{k=2}^{\infty} \frac{\sin(2k-1)\omega t}{2k-1}.$$

In the describing function analysis of the relay feedback system, the relay is replaced with its quasi-linear equivalent DF, and a self-sustained oscillation of amplitude, a and frequency, ω_{osc} is assumed. Then, if $G_p(s)$ denotes the transfer function of the process, the variables in the loop must satisfy the following relations,

$$E = -Y,$$
$$U = N(a)E,$$
$$Y = G_p(j\omega_{\text{osc}})U.$$

This implies that it must follow

$$G_p(j\omega_{\text{osc}}) = -\frac{1}{N(a)}.$$

Relay feedback estimation of the critical point for process control is thus based on the key observation that the intersection of the Nyquist curve of $G_p(j\omega)$ and $-\frac{1}{N(a)}$ in the complex plane gives the critical point of the linear process. Hence, if there is a sustained oscillation in the system of Fig. 4.3, then in the steady state, the oscillation must be at the ultimate frequency, i.e.

$$\omega_\pi = \omega_{\text{osc}},$$

and the amplitude of the oscillation is related to the ultimate gain, k_π by

$$k_\pi = \frac{4\mu}{\pi a}.$$

It may be advantageous to use a relay with hysteresis as shown in Fig. 4.4

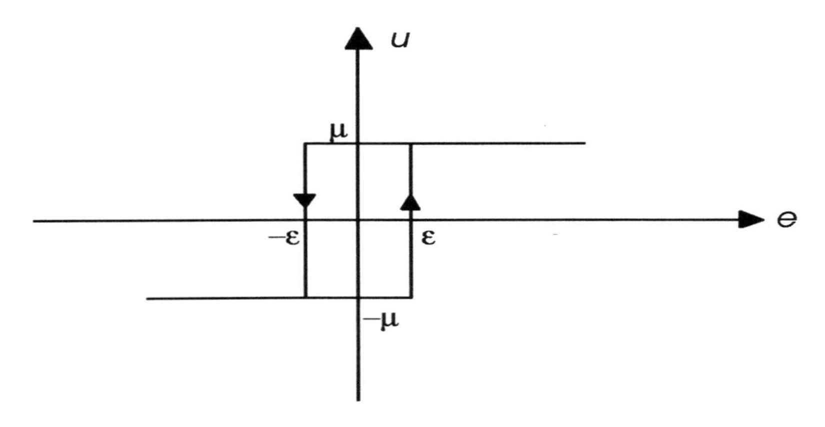

Fig. 4.4. Relay with hysteresis

so that the resultant system is less sensitive to measurement noise. The inverse negative describing function of this relay is given by $-\frac{1}{N(a)} = -\frac{\pi}{4\mu}\left(\sqrt{a^2-\epsilon^2}+j\epsilon\right)$. In this case, the oscillation corresponds to the point where the negative inverse describing function of the relay crosses the Nyquist curve of the process as shown in Fig. 4.5. With hysteresis, there is an additional parameter ϵ which can, however, be set automatically based on a pre-determination of the measurement noise level.

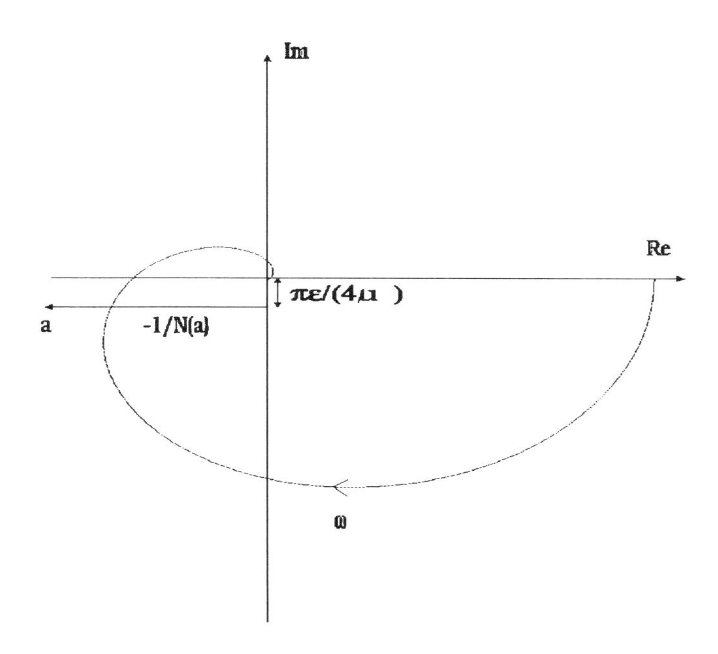

Fig. 4.5. Negative inverse describing function of the hysteretic relay.

Relay tuning is an attractively simple method for extracting the critical point of a process. Accompanying the method are three main limitations. First, the accuracy of the estimation could be poor for certain processes. Secondly, the relay experiment yields only one point of the process frequency response which may not be adequate in many applications other than raw PID tuning. Thirdly, the method is essentially an offline method which is sensitive to the presence of disturbances and it generally requires the stationary condition to be known. To overcome the limitations, some modifications to the basic method are necessary and these are covered in the subsequent sections.

4.3.2 Improved estimation accuracy

While the relay feedback experiment design will yield sufficiently accurate results for many of the processes encountered in the process control industry, there are some potential problems associated with such techniques. These arise as a result of the approximations used in the development of the procedures for estimating the critical point, i.e. the ultimate frequency and ultimate gain. In particular, the basis of most existing relay-based procedures for critical point estimation is the describing function (DF) method. This method is approximate in nature, and under certain circumstances, the existing relay-based procedures could result in estimates of the critical point that are significantly different from their real values.

The accuracy of the relay feedback estimation depends on the residual ϱ which determines whether, and to what degree, the estimation of the critical point will be successful. For the relay, ϱ consists of all the harmonics in the relay output. The amplitude of the third and fifth harmonics are about 30% and 20% that of the fundamental component, and they are not negligible if fairly accurate analysis results are desirable, and therefore they limit the class of processes for which describing function analysis is adequate, *i.e.* the process must attenuate these signals sufficiently. This is the fundamental assumption of the describing function method which is also known as the *filtering hypothesis*. Mathematically, the hypothesis requires that the process, $G_p(s)$ must satisfy

$$|G_p(jk\omega_\pi)| \ll |G_p(j\omega_\pi)| \; , \quad k = 3, 5, 7, \cdots \; , \tag{4.11}$$

and

$$|G_p(jk\omega_\pi)| \to 0 \; , \quad k \to \infty. \tag{4.12}$$

Note that (4.11) and (4.12) require the process to be not simply low-pass, but rather low-pass at the ultimate frequency. This is essential as the delay-free portion of the process may be low-pass, but the delay may still introduce higher harmonics within the bandwidth. Typical processes that fail the filtering hypothesis are processes with long time-delay and processes with resonant peaks in their frequency responses such that the undesirable frequencies are boosted instead of being attenuated. This provides one explanation for the poor results associated with these processes.

Having observed the accuracy problems associated with conventional relay feedback estimation, the design of a modified relay feedback that addresses the issue of improved estimation accuracy is considered next. Thus consider the modified relay feedback system of Fig. 4.6 where the process is assumed to have the transfer function $G_p(s)$.

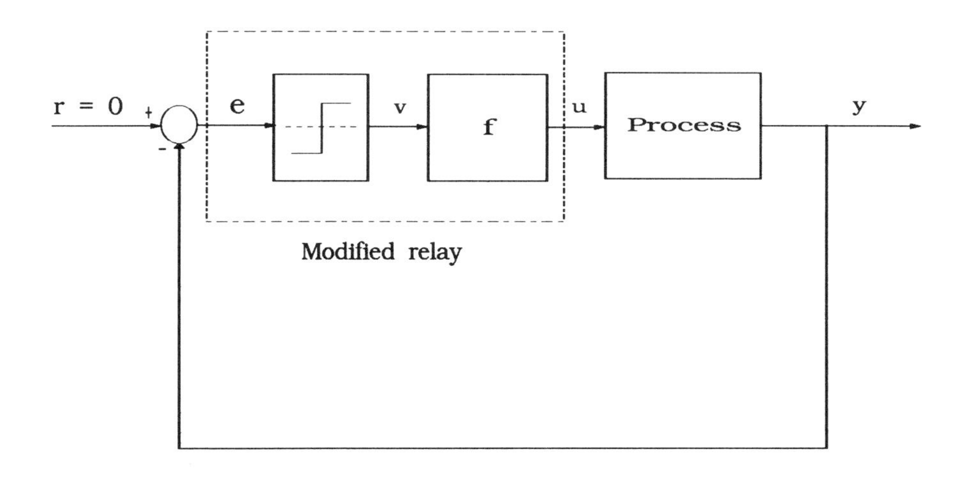

Fig. 4.6. Modified relay feedback system.

Define the mapping function f such that

$$u(t) = f(v(t)) = v_1(t) = \frac{4\mu}{\pi} \sin \omega t, \tag{4.13}$$

where $v_1(t)$ is the fundamental harmonic of $v(t)$, and μ is the amplitude of the relay element. For this modified relay feedback, it turns out that the following property may be stated:

Proposition 4.1

Consider the use of the modified relay feedback defined by the system of Fig. 4.6 and (4.13), where the process is assumed to have the transfer function $G_p(s)$. The set of signals:

$$u(t) = \frac{4\mu}{\pi} \sin(\omega^* t), \tag{4.14}$$

$$y(t) = \frac{4\mu}{\pi} A^* \sin(\omega^* t + \phi^*), \tag{4.15}$$

$$v(t) = N_r \left(-\frac{4\mu}{\pi} A^* \sin(\omega^* t + \phi^*) \right), \tag{4.16}$$

where $N_r(\cdot)$ denotes the relay function, describes an invariant set of the dynamical system defined by Fig. 4.6 for

$$\phi^* = \arg\{G_p(j\omega^*)\} = -\pi,$$
$$A^* = |G_p(j\omega^*)|.$$

Proof:

Assume that the set-up of Fig. 4.6 admits a solution of the form

$$u(t) = \frac{4\mu}{\pi} \sin(\omega^* t), \tag{4.17}$$

for some ω^*. It will be shown that this solution is consistent with the definitions of the other signals in the loop so that this solution describes an invariant set of the dynamical system defined by Fig. 4.6. With (4.17), it follows

$$y(t) = \frac{4\mu}{\pi} A^* \sin(\omega^* t + \phi^*),$$

where $A^* = |G_p(j\omega^*)|$ and $\phi^* = \arg\{G_p(j\omega^*)\}$. Then, since $e(t) = -y(t)$, it follows that

$$v(t) = N_r(e(t)) = N_r \left(-\frac{4\mu}{\pi} A^* \sin(\omega^* t + \phi^*) \right),$$

where $N_r(\cdot)$ denotes the relay function. Since this is the case, clearly $v_1(t)$, the fundamental harmonic of $v(t)$, is given by

$$v_1(t) = -\frac{4\mu}{\pi}\sin(\omega^* t + \phi^*).$$

This in turn implies that

$$u(t) = f(v(t)) = -\frac{4\mu}{\pi}\sin(\omega^* t + \phi^*). \tag{4.18}$$

(4.18) is consistent with (4.17) for

$$\phi^* = -(2n+1)\pi,$$

where n is a non-negative integer, and this thus characterizes a class of admissible solutions.

Therefore, for $n = 0$, the set of signals:

$$\begin{aligned} u(t) &= \frac{4\mu}{\pi}\sin(\omega^* t), \\ y(t) &= \frac{4\mu}{\pi}A^*\sin(\omega^* t + \phi^*), \\ v(t) &= N_r\left(-\frac{4\mu}{\pi}A^*\sin(\omega^* t + \phi^*)\right), \end{aligned}$$

describes an invariant set of the dynamical system defined by Fig. 4.6 with

$$\phi^* = -\pi,$$

as claimed. Note that the set (4.14)–(4.16) is clearly periodic in t.

Remark 4.5:

The invariant set (4.14)–(4.16) established in Proposition 4.1 provides a suitable basis for estimation of the critical point for process control with improved accuracy. This is due to the fact that the analysis used to prove Proposition 4.1 does not depend on approximations so that theoretically, it is possible to calculate the critical point exactly from observations of the invariant set (4.14)–(4.16).

To estimate the critical point using the arrangement of Fig. 4.6 and the result of Proposition 4.1, assume that an oscillation is observed which corresponds to the (admissible) solution

$$\phi^* = \arg\{G_p(j\omega^*)\} = -\pi.$$

This implies that

$$\omega^* = \omega_\pi.$$

Thus, the ultimate frequency may be obtained directly from the measurement of the frequency of the oscillation observed. For the critical point, the remaining parameter to be estimated is the ultimate gain, and from the measurements, this may be obtained from the ratio

$$k_\pi = \frac{a_\pi}{a_y} = \frac{4\mu}{\pi a_y},$$

where a_π and a_y are the observed amplitudes of the oscillations in $u(t)$ and $y(t)$ respectively. As in conventional relay feedback, the relay magnitude μ may be used as a design parameter to appropriately size the magnitude of the oscillations to handle situations with different levels of noise.

Remark 4.6:

The analysis to prove Proposition 4.1 is essentially a time-domain analysis, and no approximations were involved in establishing the existence of the invariant set (4.14)–(4.16) and the frequency ω^* that characterized it. This may be considered as an improvement over existing methods which are based on describing function analysis, and which consequently yield estimation procedures for the critical point that involve approximations. The existing methods have the advantage of simplicity, and under certain circumstances, the improved procedure here only yields small gains in accuracy. However, there are other circumstances (under-damped processes and processes with significant time-delay) where the gains in accuracy are significant.

It is interesting, actually, to note that the improved accuracy is also evident from the same describing function analysis used for the existing methods.

Thus, a describing function analysis applied to the system of Fig. 4.6 shows that corresponding to the input $e(t) = a \sin \omega t$, the non-linearity's output $u(t)$ would also be a sinusoid described by

$$u(t) = \frac{4\mu}{\pi} \sin \omega t. \tag{4.19}$$

The describing function of the non-linear system can then be obtained as

$$N(a) = \frac{4\mu}{\pi a}, \tag{4.20}$$

and it is quite straightforward to check that the residual $\varrho = 0$ so that the filtering hypothesis of the describing function analysis is satisfied strictly. It then follows that analysis using the different tools of describing functions also indicates that improved accuracy will be obtained from the procedure here.

Implementation Procedures. The system described by Fig. 4.6 and (4.13) defines the basic elements required for the improved accuracy technique. For the construction of implementation procedures, the key point to note is that the module realizing the function f should be designed to extract the fundamental harmonic, and apply it as the signal $u(t)$. There are thus various practical procedures and variations that may be used.

For a simple practical implementation, one can obtain a rough estimate of the ultimate frequency from two or three switches of the relay under normal relay feedback, and then turn on the function f using the rough estimate. The function f is then updated iteratively using the observed signals. The steps in such a procedure would be as follows:

1. Put the process under relay feedback (i.e. set $f(v(t)) = v(t)$) to obtain a rough estimate of its ultimate frequency $\tilde{\omega}_\pi$ from m oscillations, say with $m = 2$ for example.

2. Denote T_n as the time corresponding to the nth switch of the relay to $v = \mu$, and $\tilde{\omega}_{\pi,n-1}$ as the ultimate frequency estimate just prior to $t = T_n$. For $n > m$, update $f(\cdot)$ to output

$$u(t) = f(v(t)) = \frac{4\mu}{\pi} \sin \tilde{\omega}_{\pi,n-1}(t - T_n).$$

3. Obtain a new estimate of the ultimate frequency $\tilde{\omega}_{\pi,n}$ from the resultant oscillation.

4. Repeat (2)–(3) until successive estimates of the ultimate frequency show a satisfactory convergence.

Example 4.5:

The results of a simulation example utilizing this procedure for implementing f are shown in Fig. 4.7.

In the simulation, the process is given by

$$G_p(s) = \frac{1}{s+1}e^{-10s},$$

and the set-up of Fig. 4.6 is invoked. The signals $y(t)$, $u(t)$ and $v(t)$ are shown respectively in the top, middle and bottom frames in Fig. 4.7.

With the above procedure for implementing f, conventional relay feedback experiment is held from $t = 0$ to $t = 52$ (*i.e.* $m = 2$). After the third relay switch to $v = \mu$ at $t = 52$, the function $f(\cdot)$ is initialized with the ultimate frequency estimate of $\tilde{\omega}_\pi = 0.31$ obtained from the conventional experiment. Two subsequent updates of $f(\cdot)$ yield a resultant oscillation with a frequency which has essentially converged to the effectively exact ultimate frequency of $\omega_\pi = 0.29$. In the example, it can be seen that the simple method described above is relatively efficient and effective; it is also a complete and practically useful procedure for implementing the function $f(\cdot)$.

4.3.3 Estimation of a general point

With the basic relay feedback approach, only one point on the process Nyquist curve is determined. It is possible, for example, to cascade a known linear dynamical system to the system in Fig. 4.3 to obtain a frequency other than the ultimate frequency. For example, an integrator can be cascaded to obtain the point where the process Nyquist curve crosses the negative imaginary axis. Similarly, a first-order lag can be designed and cascaded to obtain a point with a frequency below the ultimate frequency. However, with these modifications, the frequency of interest cannot be specified; it is fixed by the choice of the linear element cascaded. Besides, the introduction of the linear system affects the amplitude response of the original process, and in the case of a gain reduction, a smaller signal-to-noise ratio (SNR) could affect the estimation accuracy adversely.

A variation of the basic relay will be presented in this subsection with a negative inverse describing function that is a ray through the origin in the third or fourth quadrant of the complex plane. With this particular nonlinearity, it is possible to obtain a point on the process frequency response at an arbitrarily specified phase lag. Such a point is needed, for example, in the general Ziegler-Nichols frequency response method. The system is flexible enough to give the frequency at a specified process lag of interest, $-\pi + \phi_m$ ($\phi_m \in [0, \pi)$), without affecting the amplitude response of the original process. The modified arrangement is similar to Fig. 4.6 but with f defined such that

$$u(t) = f(v(t)) = v(t - L_a(\omega)).$$

where $L_a(\omega) = \frac{\phi_m}{\omega}$, is an adaptive time delay function, and ω is the oscillating frequency of $v(t)$.

The feature will be illustrated from a describing function analysis. Consider the modified relay in the dashed-box of Figure 4.6 which consists of a relay cascaded to the delay function, f. Corresponding to the reference input $e(t) = a \sin \omega t$, the output of the modified relay can be expanded using the Fourier series, and shown to be

$$u(t) = \frac{4\mu}{\pi} \sum_{k=1}^{\infty} \frac{\sin((2k-1)\omega t - \phi_m)}{2k-1}. \tag{4.21}$$

Hence, the describing function of the modified relay can be computed as

$$N(a) = \frac{4\mu}{\pi a} e^{-j\phi_m}. \tag{4.22}$$

The negative inverse describing function, $-\frac{1}{N(a)} = \frac{\pi a}{4\mu} e^{j(-\pi+\phi_m)}$ is thus a straight line segment through the origin as shown in Fig. 4.8.

By the feedback arrangement of Fig. 4.6, the resultant amplitude and frequency of oscillation thus correspond to the intersection between $-\frac{1}{N(a)}$ and the process Nyquist curve. Hence, at the specified phase lag of $-\pi + \phi_m$, the inverse gain (k_ϕ) and the frequency of the process (ω_ϕ) can be obtained from the output amplitude (a) and frequency (ω_{osc}), i.e.

$$k_\phi = \frac{4\mu}{\pi a}, \tag{4.23}$$

$$\omega_\phi = \omega_{osc}. \tag{4.24}$$

With the arrangement of Fig. 4.6, it is thus possible to automatically track a frequency at the specified phase lag $-\pi + \phi_m$ for $\phi_m \in [0, \pi)$ without affecting the amplitude response of the original process. In this respect, it may be viewed as a generalized form of the relay feedback technique which tracks the frequency at the phase lag with $\phi_m = 0$. Besides, like relay feedback, the technique facilitates single-button tuning, a feature which is invaluable for autonomous/intelligent control applications.

Implementation Procedures. In the following, a practical implementation of procedure is presented to obtain two points on the process Nyquist curve ; $G_p(j\omega_\pi)$ at the $-\pi$ phase lag and $G_p(j\omega_\phi)$ at the specified phase lag of $-\pi + \phi_m$.

1. Estimate the process ultimate gain $k_\pi = \frac{1}{|G_p(j\omega_\pi)|}$ and frequency ω_π with the normal relay feedback.

2. With these estimates, an initial guess for the frequency ($\tilde{\omega}_\phi$) at the specified phase of $-\pi + \phi_m$ is

$$\tilde{\omega}_\phi = \frac{\pi - \phi_m}{\pi}\omega_\pi,$$

   and L_a is initialized as $L_a = \frac{\phi_m}{\tilde{\omega}_\phi}$.

3. Continue the relay experiment, and adapt the delay function f to the oscillating frequency ω_{osc}, i.e. $L_a = \frac{\phi_m}{\omega_{osc}}$. Upon convergence, $\omega_\phi = \omega_{osc}$, and $k_\phi = \frac{1}{|G_p(j\omega_\phi)|} = \frac{4\mu}{\pi a}$, where a is the amplitude of process output oscillation.

4.3.4 Estimation of multiple points

While the estimation procedure can give a general frequency of interest, tuning time is increased proportionally when more frequency estimations are required, especially if high accuracy is desirable. This is particularly true in the case of a process with a long time-delay where tuning time is considerably long. A further extension of procedure is possible which allows multiple frequency estimations in one single relay experiment. The arrangement and implementation is similar to Fig. 4.6 but with the mapping function f defined such that

$$u(t) = f(v(t)) = \sum_{k=1}^{k_m} a_k \sin k\omega t \ , \quad k_m \in Z^+,$$

where $v_1 = a_1 \sin \omega t$ is the fundamental frequency of the input v, $a_k = \frac{4\mu}{k\pi}$ and $k_m\omega$ is the upper bound of the frequencies injected. The multiple points on the frequency response can then be estimated as

$$G_p(jk\omega_\pi) = \frac{\int_{-\frac{T_\pi}{2}}^{\frac{T_\pi}{2}} y(t)e^{-jk\omega_\pi t}dt}{\int_{-\frac{T_\pi}{2}}^{\frac{T_\pi}{2}} u(t)e^{-jk\omega_\pi t}dt}, \quad k = 1 \cdots k_m, \tag{4.25}$$

where $T_\pi = \frac{2\pi}{\omega_\pi}$.

Implementation Procedures. The steps in a practical implementation of the procedure would be as follows. For good estimation accuracy from this procedure, the estimation procedure is restricted to two frequencies, i.e. $k_m = 2$.

1. Put the process under relay feedback (i.e. set $f(v(t)) = v(t)$) to obtain a rough estimate of its ultimate frequency $\tilde{\omega}_\pi$ from m oscillations, say with $m = 2$ for example.

2. Denote T_n as the time corresponding to the nth switch of the relay to $v = \mu$, and $\tilde{\omega}_{\pi,n-1}$ as the ultimate frequency estimate just prior to $t = T_n$. For $n > m$, update $f(\cdot)$ to output

$$u(t) = f(v(t)) = \frac{4\mu}{\pi} \sin \tilde{\omega}_{\pi,n-1}(t - T_n) + \frac{2\mu}{\pi} \sin 2\tilde{\omega}_{\pi,n-1}(t - T_n).$$

3. Obtain a new estimate of the ultimate frequency $\tilde{\omega}_{\pi,n}$ from the resultant oscillation.

4. Repeat (2)–(3) until successive estimates of the ultimate frequency show a satisfactory convergence. The amplitude and phase of the resultant oscillation can be obtained from (4.25).

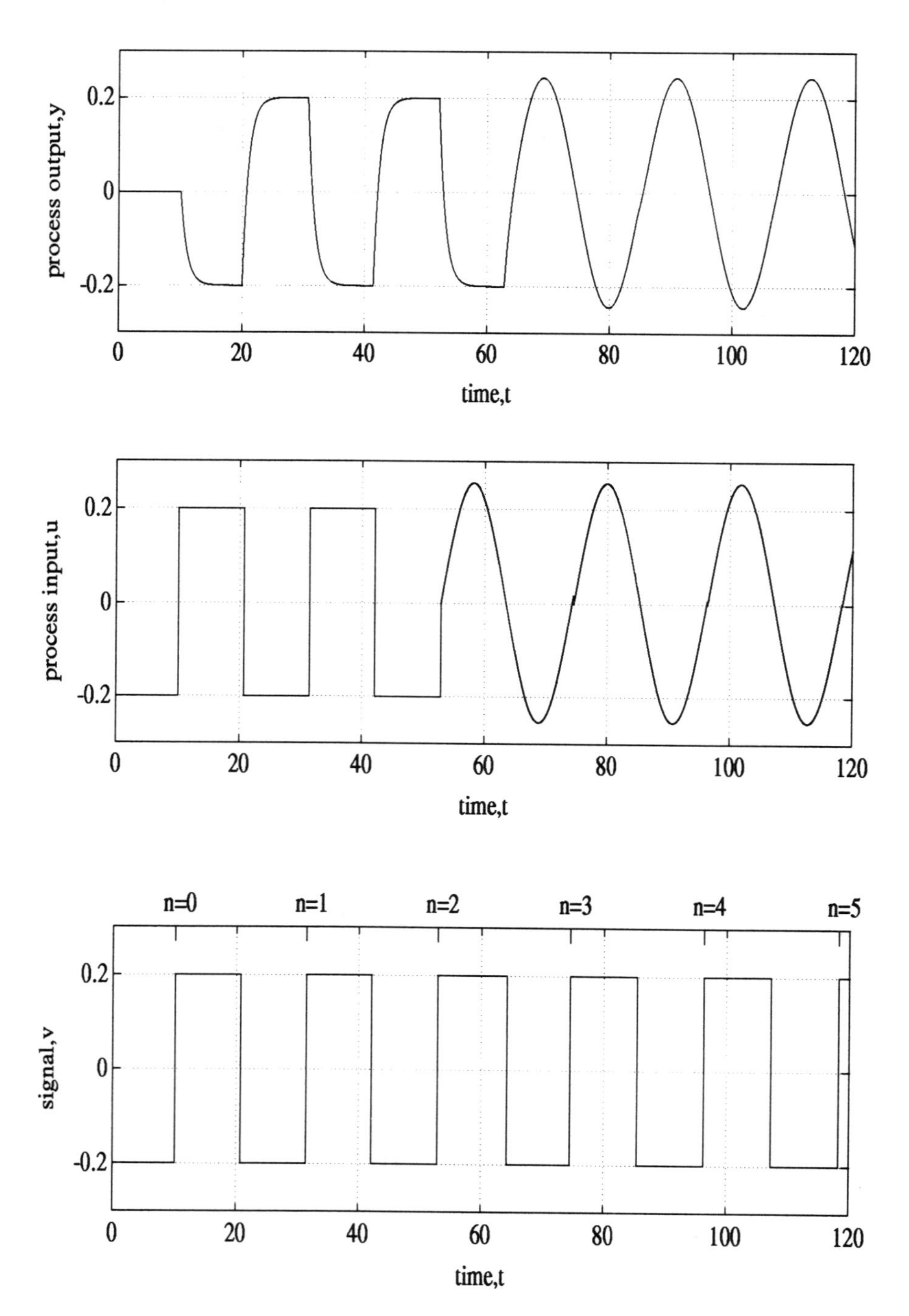

Fig. 4.7. Example for illustration of the implementation procedure.

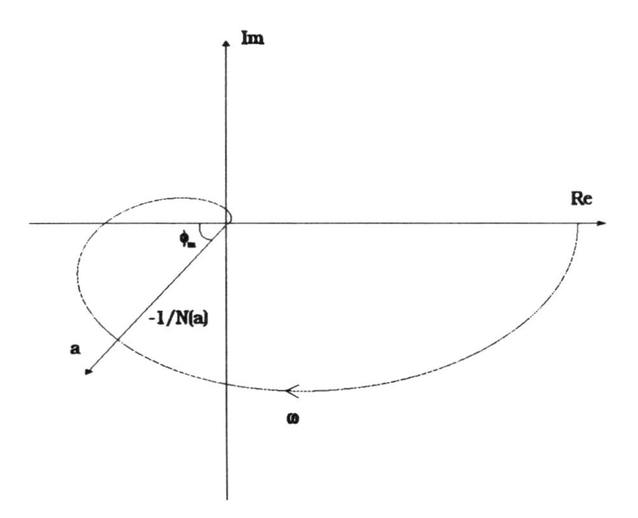

Fig. 4.8. Negative inverse describing function of the modified relay.

4.4 On-line Relay Tuning

One of the main features of the relay auto-tuning method, which probably accounts for its success more than any other associated features, is that it is a closed-loop method and therefore an on-off regulation of the process may be maintained even when the relay experiment is being conducted. However, apart from those addressed in Section 4.3.1-4.3.3, the approach has several important practical constraints related to the structure which have remained, in large proportion, unresolved to-date. First, it has a sensitivity problem in the presence of disturbance signals, which may be real process perturbation signals or equivalent ones arising from varying process dynamics, non-linearities and uncertainties present in the process. For small and constant disturbances, given that stationary conditions are known, an iterative solution has been proposed, essentially by adjusting the relay bias until symmetrical limit cycle oscillations are recovered. However, for general disturbance signals, there has been no effective solution to-date. Secondly, relating partly to the first problem, relay tuning may only begin after stationary conditions are attained in the input and output signals, so that the relay switching levels may be determined with respect to these conditions and the static gain of the process. In practice, under open-loop conditions, it is difficult to determine when these conditions are satisfied and therefore when the relay experiment may be initiated. Thirdly, the relay auto-tuning method is not applicable to certain classes of processes which are not relay-stabilizable, such as the double integrator and runaway processes. For these processes, relay feedback is not able to effectively induce stable limit cycle oscillations. Finally, the basic relay method is an off-line tuning method, i.e., some information on the process is first extracted with the process under relay feedback and detached from the controller. The information is subsequently used to commission the controller. Off-line tuning has associated implications in the tuning-control transfer, affecting operational process regulation which may not be acceptable for certain critical applications. Indeed, in certain key process control areas (e.g., vacuum control, environment control, etc) directly affecting downstream processes, it may be just too expensive or dangerous for the control loop to be broken for tuning purposes, and tuning under tight continuous closed-loop control (not the on-off type) is necessary.

In this section, the development of a robust on-line relay automatic tuning method for general control applications is considered. The on-line method is effective against many of the above-mentioned constraints of the basic relay auto-tuning method, and it retains the simplicity of the original method. In the on-line configuration, the relay is applied to an inner loop comprising of a controller-stabilized process in the usual manner. The initial settings of the controller can be conservative and intended primarily for the purpose of stabilizing the process. They may be based on simple prior information about the process or default settings may be used. Indeed, practical applications of controller automatic tuning methods have been mainly to derive more efficient updates of current or default control settings. Thus, this configuration does not strictly pose additional and stringent prerequisites for its usage, but rather, it uses information on the system which is already available in many cases.

The controller may be designed based on the methods described in Chapters 2 and 3. In this section, two new methods for online PID tuning are described - a direct or an indirect approach. In the direct approach, the control gains are tuned directly based on frequency response information obtained from the experiment so that the actual closed-loop behavior approaches the desired one. In the indirect approach, the process is approximated by a model and the parameters of the model are obtained from the signals resulting from the configuration. The controller is subsequently designed based on the process model. Both approaches will be described, in this section, with respect to the new on-line tuning configuration. In contrast to the original method, the relay switching levels are designed directly as deviations from the reference signal value, and therefore, the tuning experiment may be conducted even when the system has not lapsed into stationary conditions. Furthermore, the scope of the applications of the on-line auto-tuner is wide. It may be used in new or existing control systems, the only signals required for the interface are the process variable and the reference variable.

Simulations are provided to illustrate the applicability of the on-line approach under the various scenarios where the basic relay auto-tuning method fails to perform effectively. Practical application of the on-line approach is also demonstrated using an industrial process emulator.

4.4.1 Configuration

The configuration of the robust on-line relay auto-tuning method is given in Fig. 4.9.

Essentially, in this configuration, the relay is applied to an inner control loop comprising of the controller and the process, where the transfer function of the inner loop is given by

$$G_{yr}(s) = \frac{G_{c0}(s)G_p(s)}{1 + G_{c0}(s)G_p(s)}, \tag{4.26}$$

where $G_{c0}(s)$ denotes the PID controller during the tuning process which may be in either of the following forms:

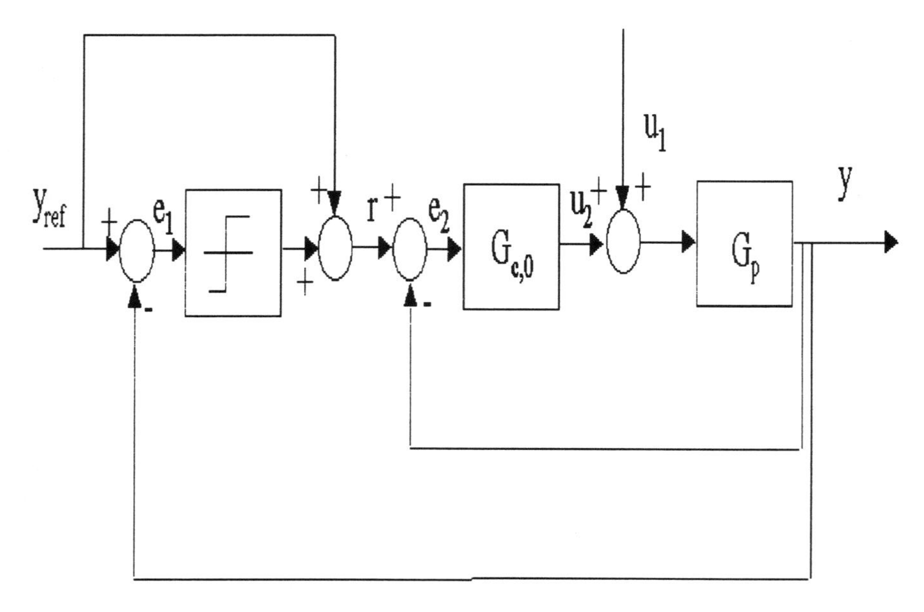

Fig. 4.9. Configuration of on-line tuning method

$$G_{c0}(s) = K_{c0}\left(1 + \frac{1}{T_{i0}s} + T_{d0}s\right),$$

$$G_{c0}(s) = K_{c0}\left(1 + \frac{1}{T_{i0}s}\right)(1 + sT_{d0})$$

where K_{c0}, T_{i0} and T_{d0} are known control gains. $G_c(s)$ will be used to denote the subsequently tuned controller and it is described by

$$G_c(s) = K_c\left(1 + \frac{1}{T_i s} + T_d s\right),$$

$$G_c(s) = K_c\left(1 + \frac{1}{T_i s}\right)(1 + sT_d)$$

where K_c, T_i and T_d are control gains to be determined from the tuning experiment.

4.4.2 Structure identification

With different structures of PID controllers present, it is sometimes necessary to be able to identify the actual structure of the controller in use and the exact controller gains in force. The former is necessary irregardless of the type of implementation, while the latter is only unnecessary for analog PID controllers, where due to normal wear and tear, it can be the situation that the indication on the knob or dial is not necessarily the actual control gains being used, and a re-calibration of the gains is necessary. In fact, experiments have shown that these deviations could be as large as 100%.

The problem of not knowing the exact functioning of the controller is not so great when the controller is tuned by trial and error methods. However, if systematic tuning methods are used, such as those described in Chapters 2 and 3, this uncertainty may result in drastically degraded control.

In this section, the application of the configuration of Fig. 4.9 for PID control structure and parameter identification will be elaborated. To this end, the following two cases are considered:

Case I: PID structure unknown; PID parameters known. If the system settles in a steady-state limit cycle oscillation at frequency ω_u, $G_c(j\omega_u)$ can be obtained from amplitude and phase response considerations in the usual manner. It then follows that multiple points of the frequency response of $G_c(s)$ may be obtained from the oscillating signals. Since the oscillations are periodic, Fourier or spectral analysis may be efficiently applied to whole cycles of the periodic signals e_2 and u_2 to yield good estimates of $G_c(j\omega)$ at odd multiples of the fundamental frequency, i.e., at $\omega = k\omega_u, k = 1, 3, 5, \ldots$.

$$G_c(jk\omega_u) = \frac{\int_0^{T_u} u_2(t)e^{-jk\omega_u t}dt}{\int_0^{T_u} e_2(t)e^{-jk\omega_u t}dt}, k = 1, 3, 5, ..., \tag{4.27}$$

where $T_u = \frac{2\pi}{\omega_u}$ is the period of the stationary oscillations. If there are 2^n samples, where n is an integer, the *Fourier Transforms* can be efficiently computed using the *Fast Fourier Transform* (FFT) algorithm as

$$G_c(jk\omega_u) = \frac{FFT(u_2)}{FFT(e_2)}. \tag{4.28}$$

Since the signals concerned here are periodic at steady state, the empirical transfer function estimates thus obtained will exhibit good statistical properties, and they will be robust to noise if sufficiently large number of samples are used relative to the signal-noise-ratio (SNR). This is the main appeal of FFT which has resulted in it being one of the most practically useful tools in control system analysis and design.

The approach here to identify PID structure is to compute the PID parameters corresponding to the different structures considered and construct the parallel and series PID controller with two sets of value. Then with the known information of $G_c(jk\omega_u)$, the right structure can be identified. The parameter estimation formulas are given below for the two configurations under consideration.

Parallel type

Assuming the PID controller is of the parallel type, the frequency response of the controller may be obtained at these points as

$$\tilde{G}_c(jk\omega_u) = K_{cp}\left(1 + \frac{1}{jT_{ip}k\omega_u} + jT_{dp}k\omega_u\right), \quad k = 1, 3, 5, ..., 2n-1, \tag{4.29}$$

where K_{cp}, T_{ip} and T_{dp} are the control parameters to be determined. This may be re-written as:

$$\tilde{G}_c(jk\omega_u) = K_{cp} + j\left(K_{cp}T_{dp}k\omega_u - \frac{K_{cp}}{T_{ip}k\omega_u}\right), \quad k = 1, 3, 5, ..., 2n-1. \tag{4.30}$$

Define, $\epsilon(\omega)$, a measure of the mismatch between $G_c(j\omega)$ and $\tilde{G}_c(j\omega)$ as

$$\epsilon(\omega) = G_c(j\omega) - \tilde{G}_c(j\omega). \tag{4.31}$$

A suitable loss function is then given by:

$$J = \sum_k |\epsilon(k\omega_u)|^2, \quad k = 1, 3, 5, ..., 2n-1. \tag{4.32}$$

A direct way of determining the unknown structure of the controller will be to compute the loss functions J_p and J_s corresponding to the parallel and series structure with the known parameters. The structure yielding the smaller J will be identified as the correct structure. An alternative method would be to compute the PID parameters for both structures and compare these parameters with the known values. The latter method is useful when the known

PID parameters are inaccurate, especially for analog PID controllers. This method yields the structure as well as the actual PID parameters in force.

The PID parameter set will be computed so that $\epsilon(\omega)$ is a minimum in the least squares sense. Minimising $Re(\epsilon(j\omega))$ and $Im(\epsilon(j\omega))$ respectively, it follows that:

$$K_{cp} = \frac{\sum_k Re(G_c(jk\omega_u))}{n}, \quad k = 1, 3, 5, ..., 2n-1. \tag{4.33}$$

and

$$\begin{bmatrix} \omega & \frac{1}{\omega} \\ \vdots & \vdots \\ k\omega & \frac{1}{k\omega} \\ \vdots & \vdots \\ (2n-1)\omega & \frac{1}{(2n-1)\omega} \end{bmatrix} \begin{bmatrix} K_{cp}T_{dp} \\ -\frac{K_{cp}}{T_{ip}} \end{bmatrix} = \begin{bmatrix} Im(G_c(j\omega)) \\ \vdots \\ Im(G_c(jk\omega)) \\ \vdots \\ Im(G_c(j(2n-1)\omega)) \end{bmatrix}, \quad k = 1, 3, 5, ..., 2n-1. \tag{4.34}$$

(4.34) has been posed in the linear regression form. Least squares solutions for the remaining two parameters may be computed non-iteratively as:

$$T_{ip} = \frac{K_{cp}(\sum_k k^2\omega_u^2 \sum_k \frac{1}{k^2\omega_u^2} - n^2)}{-\sum_k k^2\omega_u^2 \sum_k \frac{1}{k\omega_u} Im(G_c(jk\omega_u)) + n\sum_k k\omega_u Im(G_c(jk\omega_u))}, \quad k = 1, 3, 5, ..., 2n-1. \tag{4.35}$$

and

$$T_{dp} = \frac{\sum_k \frac{1}{k^2\omega_u^2} \sum_k k\omega_u Im(G_c(jk\omega_u)) - n\sum_k \frac{1}{k\omega_u} Im(G_c(jk\omega_u))}{K_{cp}(\sum_k k^2\omega_u^2 \sum_k \frac{1}{k^2\omega_u^2} - n^2)}, \quad k = 1, 3, 5..., 2n-1. \tag{4.36}$$

In practice, most systems are low-pass in nature, so that it is reasonable to fix an upper bound for k at empirically $k = 5$, i.e., $n = 3$.

Series type

The frequency response of a series type PID controller at odd multiples of ω_u is given by:

$$\tilde{G}_c(jk\omega_u) = K_{cs}\left(1 + \frac{1}{jT_{is}k\omega_u}\right)(1 + jT_{ds}k\omega_u), \quad k = 1, 3, 5, ..., 2n-1. \tag{4.37}$$

This may be re-written as

$$\tilde{G}_c(jk\omega_u) = K_{cs}\left(1 + \frac{T_{ds}}{T_{is}}\right) + j\left(K_{cs}T_{ds}k\omega_u - \frac{K_{cs}}{T_{is}k\omega_u}\right), \quad k = 1, 3, 5, ..., 2n-1 \tag{4.38}$$

Following the same computations as for the parallel type of PID control, the various control parameters are computed as follow:

$$K_{cs} = \frac{\theta_2 + \sqrt{\theta_2^2 + 4\theta_1}}{2}, \tag{4.39}$$

$$T_{is} = \frac{K_{cs}(\sum_k k^2\omega_u^2 \sum_k \frac{1}{k^2\omega_u^2} - n^2)}{-\sum_k k^2\omega_u^2 \sum_k \frac{1}{k\omega_u} Im(G_c(jk\omega_u)) + n\sum_k k\omega_u Im(G_c(jk\omega_u))}, \quad k = 1, 3, 5, ..., 2n-1. \tag{4.40}$$

and

$$T_{ds} = \frac{\sum_k \frac{1}{k^2\omega_u^2} \sum_k k\omega_u Im(G_c(jk\omega_u)) - n\sum_k \frac{1}{k\omega_u} Im(G_c(jk\omega_u))}{K_{cs}(\sum_k k^2\omega_u^2 \sum_k \frac{1}{k^2\omega_u^2} - n^2)}, \quad k = 1, 3, 5, ..., 2n-1. \tag{4.41}$$

where

$$\theta_1 = (-\sum_k k^2\omega_u^2 \sum_k \frac{1}{k\omega_u} Im(G_c(jk\omega_u)) + n\sum_k k\omega_u Im(G_c(jk\omega_u))$$
$$(\sum_k \frac{1}{k^2\omega_u^2} \sum_k k\omega_u Im(G_c(jk\omega_u))$$
$$-n\sum_k \frac{1}{k\omega_u} Im(G_c(jk\omega_u))/(\sum_k k^2\omega_u^2 \sum_k \frac{1}{k^2\omega_u^2} - n^2)^2,$$
$$k = 1, 3, 5, ..., 2n-1. \tag{4.42}$$

$$\theta_2 = \frac{\sum_k Re(G_c(jk\omega_u))}{n}, \; k = 1, 3, 5, ..., 2n-1. \tag{4.43}$$

Case II: Both PID parameters and structure unknown. In certain scenarios, both parameters and structure of the PID controller may be unknown which makes the identification more difficult. For analog version of PID controllers, even if the parameters are known, they often become largely inaccurate with time due to aging and wear of components. In these cases, it is still necessary to first identify the parameters before the structure.

In this section, a systematic method is developed to obtain both the parameters and structure of controller in two phases of the experiment. In the initial phase, the derivative part is disabled, and the other two parameters are identified in the usual way. The remaining unknown D parameter and the controller structure are subsequently obtained in the second phase of the experiment.

Phase I: Identification of K_c and T_i

With the D part disabled ($T_d = 0$), both parallel and series types are similar in structure. This common structure is referred to as:

$$G_c(s) = K_c\left(1 + \frac{1}{T_i s}\right). \tag{4.44}$$

K_c and T_i are obtained in the same way as described in Case I.

$$K_c = \frac{\sum_k Re(G_c(jk\omega_u))}{n}, \; k = 1, 3, 5, ..., 2n-1. \tag{4.45}$$

$$T_i = -\frac{K_c \sum_k \frac{1}{k^2\omega_u^2}}{\sum_k \frac{1}{k\omega_u} Im(G_c(jk\omega_u))}, \; k = 1, 3, 5, ..., 2n-1. \tag{4.46}$$

Phase II: Identification of T_d and control structure

After K_c and T_i are identified, the D part is reconnected and the experiment repeated. This phase may be divided into two parts.

Computation of T_d

The main idea here is based on the observation that the frequency response of both the parallel and series types of PID share the same imaginary portion (see (4.30) and (4.38)). Using the same definition of $\epsilon(\omega) = G_c(j\omega) - \tilde{G}_c(j\omega)$, T_d may thus be computed such that $Im(\epsilon(\omega))$ is a minimum in the least squares sense. It may be directly shown that T_d can be estimated as

$$T_d = \frac{\sum_k k\omega_u (Im(G_c(jk\omega_u)) + \frac{K_c}{T_i k\omega_u})}{K_c \sum_k k^2\omega_u^2}, \; k = 1, 3, 5, ..., 2n-1. \tag{4.47}$$

Identification of structure

At this point of experimentation, K_c, T_i and T_d of the PID controller are all known. The controller can be constructed in parallel and series form with these parameters. Using the same measure of the total residual errors as in (4.32), J may be computed for both the parallel and series type structure and the structure with the smaller J is identified as the controller structure.

4.4.3 Assessment of Control Performance

In this section, guidelines are presented for the evaluation of control performance against user specifications, from the same information available from the on-line identification experiment.

Benchmark against frequency response prototype. If a desired frequency response prototype (i.e., $\bar{G}_{yr}(j\omega)$) is available, assessment of control performance in terms of speed and stability can be made with reference to this specification.

A suitable stability measure is

$$\beta = \frac{|G_{yr}(j\omega_u)|}{|\bar{G}_{yr}(j\omega_u)|}. \tag{4.48}$$

Typically, the stability of the closed-loop system will satisfy the specification if $\beta \leq 1$.

A suitable measure of speed is:

$$\gamma = [(arg(\bar{G}_{yr}(j\omega_u) - arg(G_{yr}(j\omega_u))]/\pi. \tag{4.49}$$

Note that, typically the speed requirement is met if $\gamma \geq 0$.

These guidelines will be used in the on-line assessment of PID controllers in the following sections.

4.4.4 Controller design

In this section, two simple and effective tuning methods are presented for PID controllers, whichever structure it takes. The first is a direct tuning method based on desired frequency response prototype, $\bar{G}_{yr}$. The second is an indirect method which first yields a process model from the on-line relay experiment. Many methods are available to subsequently tune PID controllers based on the explicit process model.

Direct method.

The direct method will be illustrated for both the parallel and series types of PID control structures.

Parallel type

If the system settles in a steady-state limit cycle oscillation at frequency ω_u, then $G_{yr}(j\omega_u)$ can be obtained from amplitude and phase response considerations in the usual manner. It then follows that one point of the frequency response of $G_p(j\omega)$ may be obtained at $\omega = \omega_u$ as:

$$G_p(j\omega_u) = \frac{G_{yr}(j\omega_u)}{G_c(j\omega_u)(1 - G_{yr}(j\omega_u))}. \tag{4.50}$$

Let the frequency response prototype at the frequency ω_u be denoted by: $\bar{G}_{yr}(j\omega_u)$. The desired frequency response of the controller at this frequency to achieve this specification is thus given by

$$\bar{G}_c(j\omega_u) = \frac{\bar{G}_{yr}(j\omega_u)}{G_p(j\omega_u)(1 - \bar{G}_{yr}(j\omega_u))}. \tag{4.51}$$

The new gains K_{cp}, T_{ip} and T_{pd} of the controller $G_c(s)$ may thus be chosen so that $G_c(j\omega) = \bar{G}_c(j\omega)$ for $\omega = \omega_u$. Note that there is one more unknown (3) than the number of real equations (2). Thus, in addition, T_{ip} and T_{dp} may be chosen subject to the heuristic rule $T_{dp} = 0.25T_{ip}$ as suggested by Astrom and Hagglund (1995).

The desired controller frequency response $\bar{G}_c(j\omega_u)$ may then be derived from $G_{yr}(j\omega_u)$ and $\bar{G}_{yr}(j\omega_u)$ as:

$$\bar{G}_c(j\omega_u) = \frac{\bar{G}_{yr}(j\omega_u)(1 - G_{yr}(j\omega_u))G_{c0}(j\omega_u)}{G_{yr}(j\omega_u)(1 - \bar{G}_{yr}(j\omega_u))}. \tag{4.52}$$

Using the PID controller with $T_{dp} = 0.25T_{ip}$, $G_c(j\omega)$ may be written as:

$$G_c(j\omega_u) = \frac{K_{cp}(0.5T_{ip}j\omega_u + 1)^2}{T_{ip}j\omega_u}. \tag{4.53}$$

Matching $G_c(j\omega_u)$ and $\bar{G}_c(j\omega_u)$ it can be shown that

$$T_{ip} = \frac{2}{\omega_u} \tan \frac{\pi/2 + arg(\bar{G}_c(j\omega_u))}{2}, \tag{4.54}$$

and

$$K_{cp} = \frac{T_{ip}\omega_u|\bar{G}_c(j\omega_u)|}{0.25T_{ip}^2\omega_u^2 + 1}. \tag{4.55}$$

Series type

The parameters of the series type may be obtained following the simliar procedures as the parallel type. Choosing $T_{ds} = 0.25T_{is}$, the parameters are

$$T_{is} = \frac{\sqrt{6.25{\omega_u}^2 + 4{\omega_u}^2\tan^2\frac{2\arg(\bar{G}_c(j\omega_u))+\pi}{2}} - 2.5\omega_u}{\omega_u^2\tan\frac{2\arg(\bar{G}_c(j\omega_u))+\pi}{2}}, \tag{4.56}$$

and

$$K_{cs} = \frac{T_{is}\omega_u|\bar{G}_c(j\omega_u)|}{\sqrt{T_{is}^2\omega_u^2 + 1}\sqrt{0.0625T_{is}^2\omega_u^2 + 1}}. \tag{4.57}$$

Indirect method.

In the indirect method, the process is first fitted to a transfer function model using parameter estimation approaches on the relay-induced oscillation signals. The first order with dead-time transfer function is again assumed.

$$\tilde{G}_p(s) = \frac{K_p}{1+Ts}e^{-sL} \tag{4.58}$$

Based on this model, the controller may be designed. With reference to Fig. 4.10, if u_2 and y are measurable, the parameters of $\tilde{G}_p(s)$ may be directly computed. If only r and y are measurable, the following formulas may be used instead.

Parallel type

From (4.51) and (4.52),

$$arg[G_c(j\omega_u)G_p(j\omega_u)] = -\pi, \tag{4.59}$$

$$|G_c(j\omega_u)G_p(j\omega_u)| = \frac{1}{k_u+1}. \tag{4.60}$$

With G_c of the parallel type, it may be directly shown that

$$G_c(s)\tilde{G}_p(s) = \frac{K_{cp}K_p(1+sT_{ip}+s^2T_{ip}T_{dp})}{sT_{ip}(1+sT)}e^{-sL}. \tag{4.61}$$

Matching $G_c(j\omega)\tilde{G}_p(j\omega)$ to $G_c(j\omega)G_p(j\omega)$ at the frequency $\omega = \omega_u$, it follows:

$$\frac{1}{2}\pi + \arctan\frac{\omega_u}{\alpha_1} + \arctan\frac{\omega_u}{\alpha_2} - \arctan\omega_u T - \omega_u L = 0, \tag{4.62}$$

$$(k_u+1)K_{cp}K_p = \omega_u T_i\sqrt{\frac{\omega_u^2T^2+1}{(\omega_u^2+\alpha_1^2)(\omega_u^2+\alpha_2^2)}}, \tag{4.63}$$

where $\alpha_1, \alpha_2 = \frac{T_{ip}\pm\sqrt{T_{ip}^2-4T_{ip}T_{dp}}}{2T_{ip}T_{dp}}$.

Thus T and L may be obtained as:

$$T = \frac{K_{cp}K_p(k_u+1)\sqrt{(\omega_u^2+\alpha_1^2)(\omega_u^2+\alpha_2^2)}}{\omega_u^2T_{ip}}, \tag{4.64}$$

$$L = \frac{\frac{1}{2}\pi + \arctan\frac{\omega_u}{\alpha_1} + \arctan\frac{\omega_u}{\alpha_2} - \arctan\omega_u T}{\omega_u}. \tag{4.65}$$

The process steady state gain K_p is assumed to be known, or it can be obtained from a separate step experiment or a static Fourier analysis on the process input and output signals as:

$$K_p = \frac{\int_t^{t+T_u} y(t)dt}{\int_t^{t+T_u} r(t)dt}. \tag{4.66}$$

Series type

For the series type structure, (4.61) can be re-written as:

$$G_c(s)\tilde{G}_p(s) = \frac{K_{cs}K_p(1+sT_i)(1+sT_{ds})}{sT_{is}(1+sT)}e^{-sL}. \tag{4.67}$$

Matching $G_c(j\omega)\tilde{G}_p(j\omega)$ to $G_c(j\omega)G_p(j\omega)$ at the frequency $\omega = \omega_u$, it follows:

$$\frac{1}{2}\pi + \arctan\omega_u T_{is} + \arctan\omega_u T_{ds} - \arctan\omega_u T - \omega_u L = 0, \tag{4.68}$$

$$(k_u+1)K_{cs}K_p = \omega_u T_{is}\sqrt{\frac{\omega_u^2 T^2 + 1}{(\omega_u^2 T_{is}^2 + 1)(\omega_u^2 T_{ds}^2 + 1)}}. \tag{4.69}$$

Thus, T and L may be obtained as:

$$T = K_{cs}K_p(k_u+1)T_{ds}, \tag{4.70}$$

$$L = \frac{\frac{1}{2}\pi + \arctan\omega_u T_{is} + \arctan\omega_u T_{ds} - \arctan\omega_u T}{\omega_u} \tag{4.71}$$

Simulation study. In this section, the practical appeal and other useful characteristics of the new auto-tuner will be demonstrated, using scenarios where the basic relay auto-tuner is unable to perform effectively. Where possible, performance comparison between the on-line and the basic tuning method will be based on processes reported in relevant literature.

Example 4.6: Controller auto-tuning under load disturbances

The basic relay auto-tuner is unable to operate effectively under the influence of a load disturbances. For step disturbance of a small magnitude, the consequence is asymmetry in limit cycle oscillation which will typically lead to errors in frequency response estimation. For disturbance of larger magnitude, the oscillations may become unstable. An iterative relay bias adjustment method has been proposed by Hang (1993) for small step disturbances. However, apart from being an iterative approach, the method may be used only if the stationary conditions of the process are known. This example will illustrate PID controller auto-tuning with the on-line method under a load disturbance.

Consider the process evaluated by Hang (1993):

$$G_p(s) = \frac{1}{(s+1)^2} e^{-s}. \tag{4.72}$$

Both the basic and new relay auto-tuner are applied to the process at $t = 0$ and the essential signals are logged in Fig. 4.10 and Fig. 4.11 respectively. At $t = 20$, a step load disturbance seeps into the process. Note that the limit cycle oscillations with the basic method becomes asymmetrical while the new method regains the symmetry automatically after an initial transient following the occurrence of the disturbance signal. Note that the knowledge of the stationary point is neither assumed nor necessary. A new set point change follows at $t = 60$. In the basic relay method, there will be a necessity for a switch-over from relay tuning mode to closed-loop control. In the new method, no switchover is necessary since the tuning experiment is on-line, i.e., carried out under closed-loop conditions.
The control performance of the PID controller tuned using the two configurations, but using the same direct control design method are shown in Fig. 4.10 and 4.11, where it is self-evident that the on-line tuning method registers an enhanced control performance.

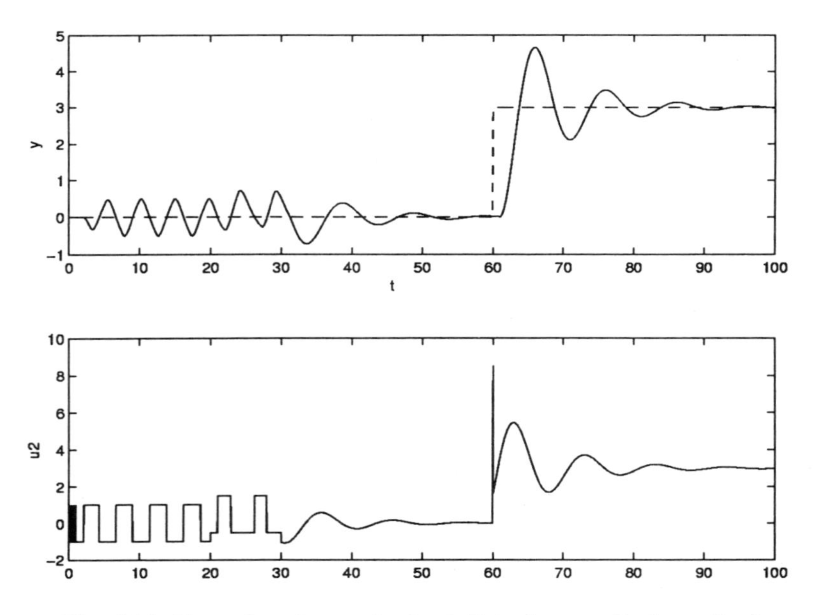

Fig. 4.10. Control tuning under load disturbance - Basic method

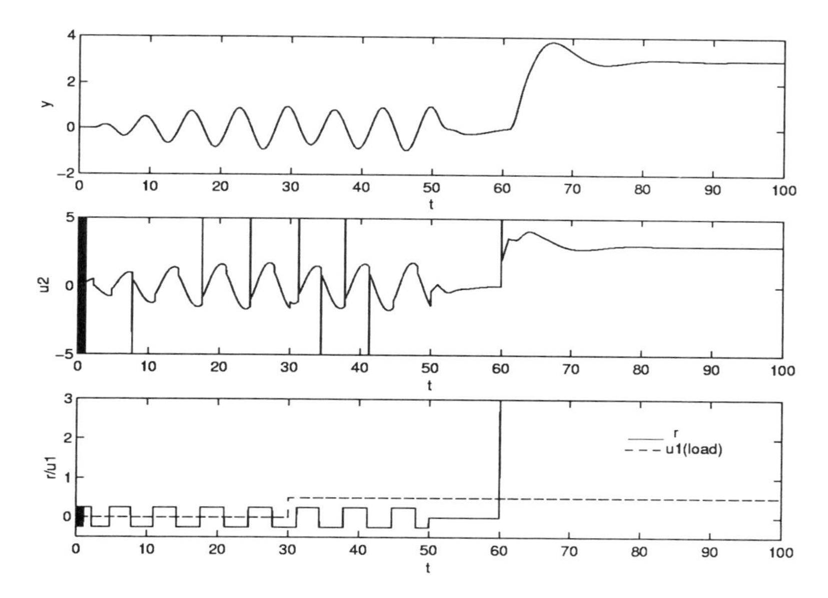

Fig. 4.11. Control tuning under load disturbance - On-line method

Example 4.7: Controller auto-tuning for time-varying processes

In this example, the robustness of the on-line controller tuning method will be illustrated when the process parameters vary during the tuning experiment. This phenomenon is quite commonly encountered in practice due to drift in instrumentation and sensing characteristics, or a change in ambient conditions. Consider the following process:

$$G_p(s) = \frac{1}{(s+1)^2} e^{-s}. \tag{4.73}$$

The initial stationary condition is assumed to be $u_0 = 0$, $y_0 = 0$ for the purpose of simulation. Unlike the basic relay tuning method, the on-line method does not need to know this condition. During the tuning experiment at $t = 14$, the process static gain changes 100% from $K_p = 1$ to $K_p = 2$. However, as illustrated in Fig. 4.12, the transient due to this abrupt parameter change is quenched in due time and stationary limit cycle oscillations re-gained. The indirect control tuning method is applied which yields the process model as

$$\tilde{G}_p(s) = \frac{2}{1 + 1.778s} e^{-1.5007s}. \tag{4.74}$$

The optimal PI tuning method via the LQR approach (Chapter 3) is applied based on this model. With the controller accordingly commissioned, a set point change occurs at $t = 40$, and the good subsequent control performance is shown in Fig. 4.12.

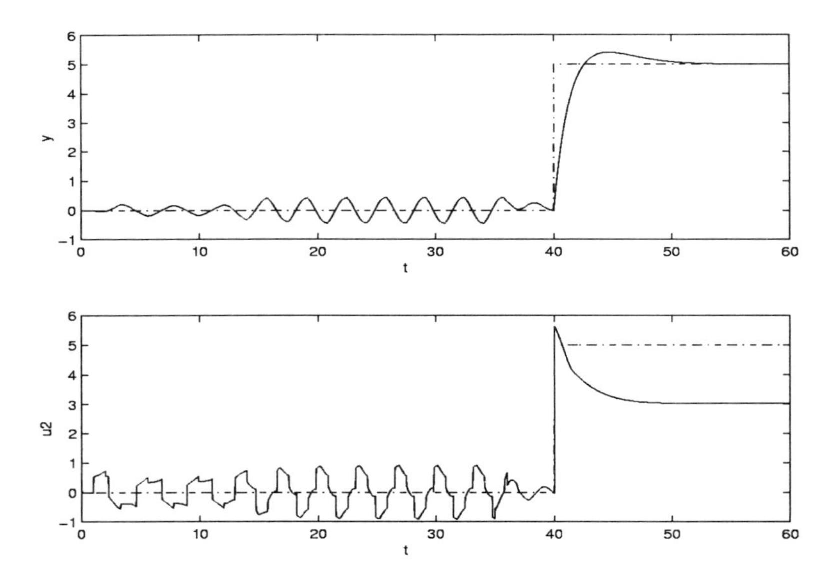

Fig. 4.12. Control tuning for a time-varying system

Example 4.8: Controller tuning under set point changes

A prominent feature and advantage of the online tuning method is that tuning may be carried out under closed-loop control. In production processes, this may mean tremendous cost savings as the production process is not disrupted during controller tuning and re-tuning.

Consider the process described by:

$$G_p(s) = \frac{1}{(s+1)^3} e^{-0.5s}. \tag{4.75}$$

The reference signal changes value during the relay experiment at $t = 25$ as shown in Fig. 4.13. Note that limit cycle oscillation continues to be stable during the duration when the process variable regulates to the new set-point.

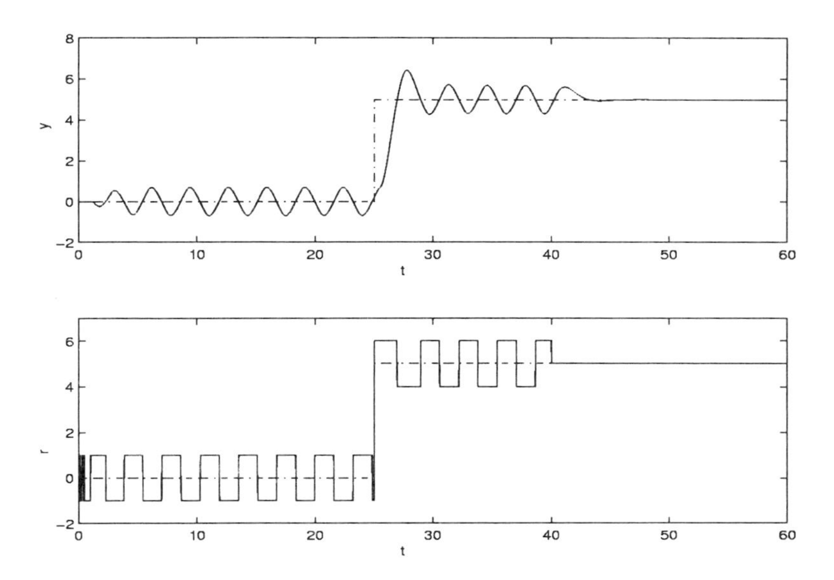

Fig. 4.13. Control tuning under set point change

Example 4.9: Controller auto-tuning for a double integrator

It is well-known that the basic relay tuning method is not applicable to the double integrator $G_p(s) = \frac{1}{s^2}e^{-0.1s}$, as it yields unstable limit cycle oscillations which increase in amplitude beyond bound as shown in Fig. 4.14. The double integrator is a feature present in several kinds of systems, one of which is the classical ball on the beam apparatus. In this example, the on-line relay tuning method is applied to a double integrator without any prior knowledge assumed of the double integrating dynamics. Stable limit cycle oscillations are attained as shown in Fig. 4.15 from which the PID controller is tuned using the direct approach presented. The subsequent PID control performance in response to a set point change is given in Fig. 4.15.

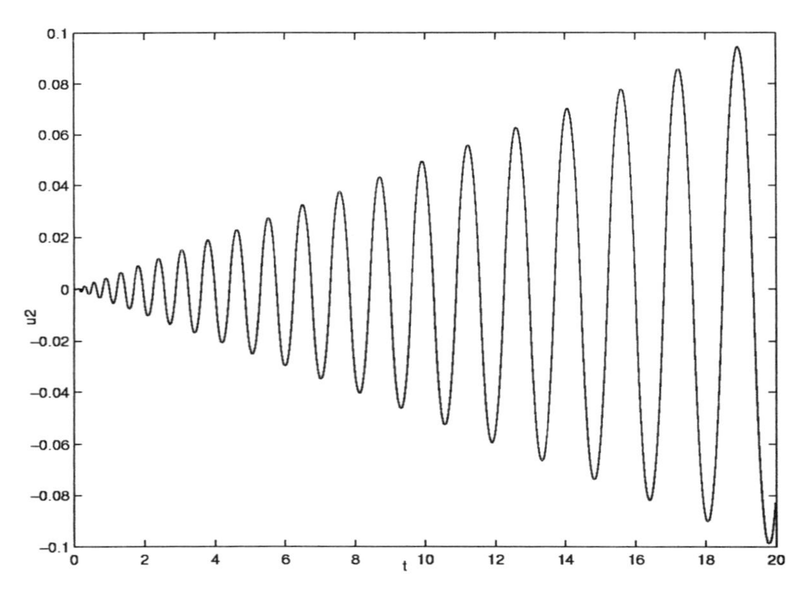

Fig. 4.14. Typical unstable limit cycle oscillation from a double integrator under relay feedback

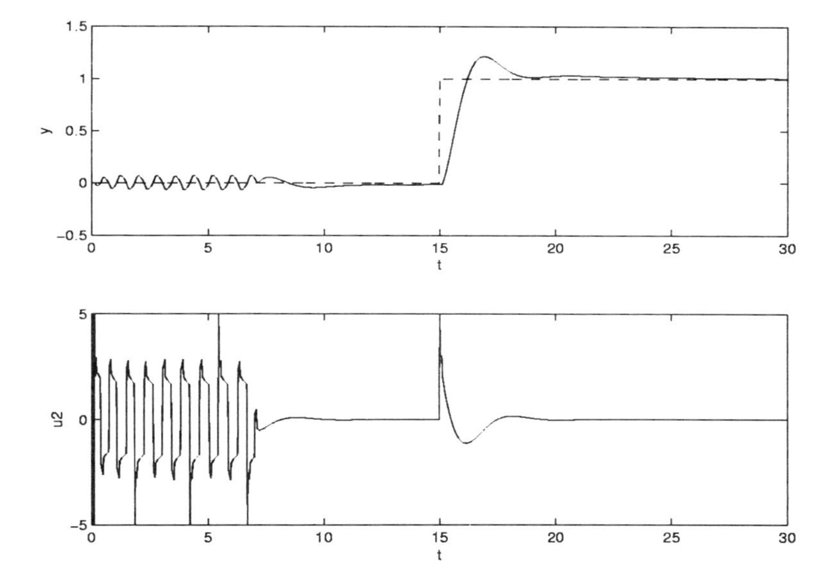

Fig. 4.15. Control tuning for a double integrator

Example 4.10: Controller auto-tuning for servo-mechanical systems

Servo-mechanical systems represent another category of systems for which the basic relay tuning method has not been effective without considering additional dynamical elements cascaded to the original system. The on-line tuning method is more effective with these systems due to the additional order of dynamics introduced by the controller. Consider a d.c. motor with velocity dynamics described by:

$$G_p(s) = \frac{4}{s^2 + s + 4} e^{-0.02s}. \tag{4.76}$$

Indirect control tuning with the on-line configuration yields the model:

$$\tilde{G}_p(s) = \frac{1}{1 + 0.14s} e^{-0.22s}. \tag{4.77}$$

The control tuning method described as in (Tan et al., 1998) is applied and the control performance achieved is highlighted in Fig. 4.16.

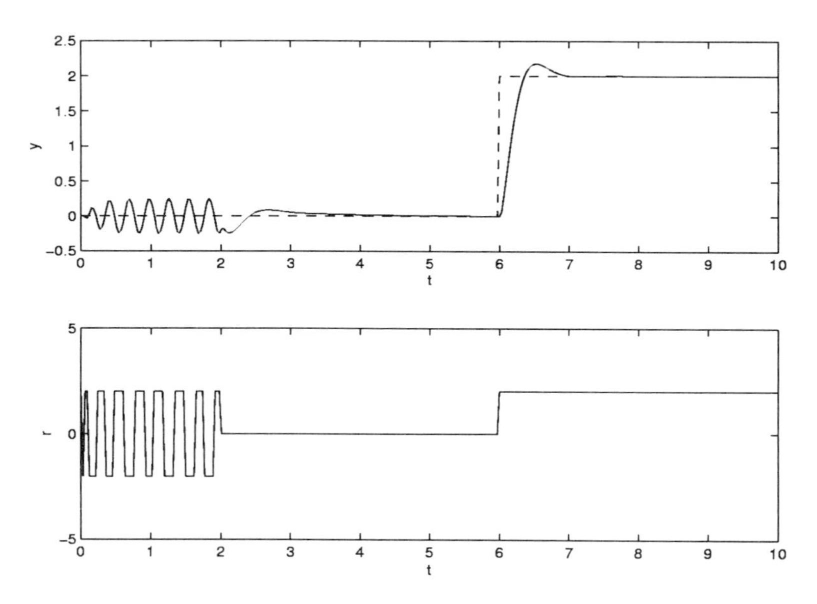

Fig. 4.16. Control tuning for a servo system

Example 4.11: Structure Identification - PID structure unknown; PID parameters known

Consider the process:

$$G_p(s) = \frac{1}{(s+1)^2}e^{-s} \tag{4.78}$$

Assume also the following parallel type PID controller has already been commissioned and is in force:

$$G_c(s) = 1.2\left(1 + \frac{1}{2.4s} + 0.9167s\right). \tag{4.79}$$

Assume further the PID parameters are known, but the structure of this controller is to be determined. From the on-line relay experiment, the input and output signals of the controller are obtained and shown in Fig. 4.17. If the parallel type is assumed, (4.33)-(4.36) yield the PID parameters as $K_{cp} = 1.22$, $T_{ip} = 2.43$ and $T_{dp} = 0.91$. If the series type is assumed, (4.39)-(4.41) yield $K_{cs} = 1.67$, $T_{is} = 3.33$ and $T_{ds} = 0.65$. Thus, the parallel structure is correctly identified.

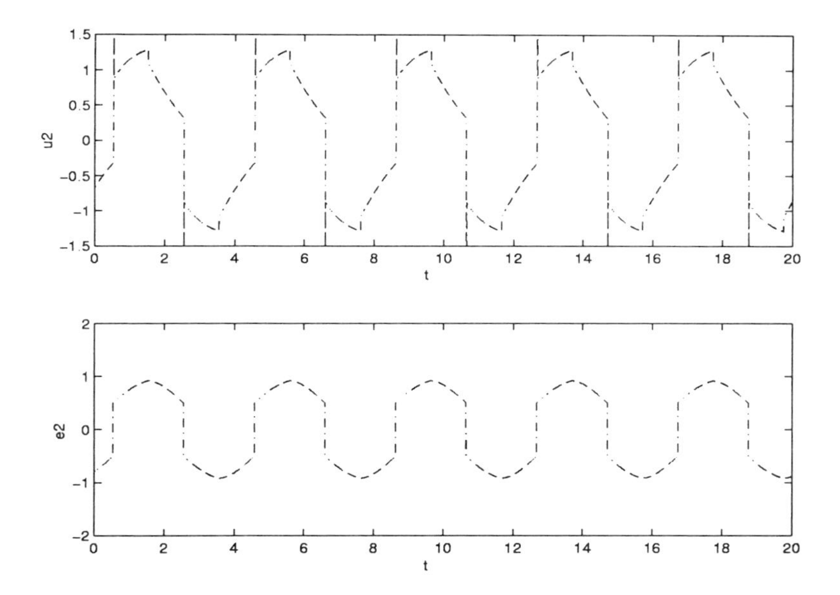

Fig. 4.17. On-line identification experiment: Unknown structure only

Example 4.12: Identification under noisy measurements

Consider the same process:

$$G_p(s) = \frac{1}{(s+1)^2}e^{-s} \tag{4.80}$$

under closed-loop control using the following series type PID controller:

$$G_c(s) = 1.2\left(1 + \frac{1}{2.1s}\right)(1 + 0.7s). \tag{4.81}$$

The signals obtained from on-line relay experiment are shown in Fig. 4.18. White noise with power density of 0.1 is added to the process measurements to simulate the effects of noisy measurements. Assuming the PID structure is of the parallel type, it follows from (4.33)-(4.36), $K_{cp} = 1.56$, $T_{ip} = 3.00$ and $T_{dp} = 0.49$. Assuming the series configuration instead, it follows from (4.39)-(4.41), $K_{cs} = 1.24$, $T_{is} = 2.38$ and $T_{ds} = 0.62$. The series configuration is thus correctly identified despite the noisy measurements.

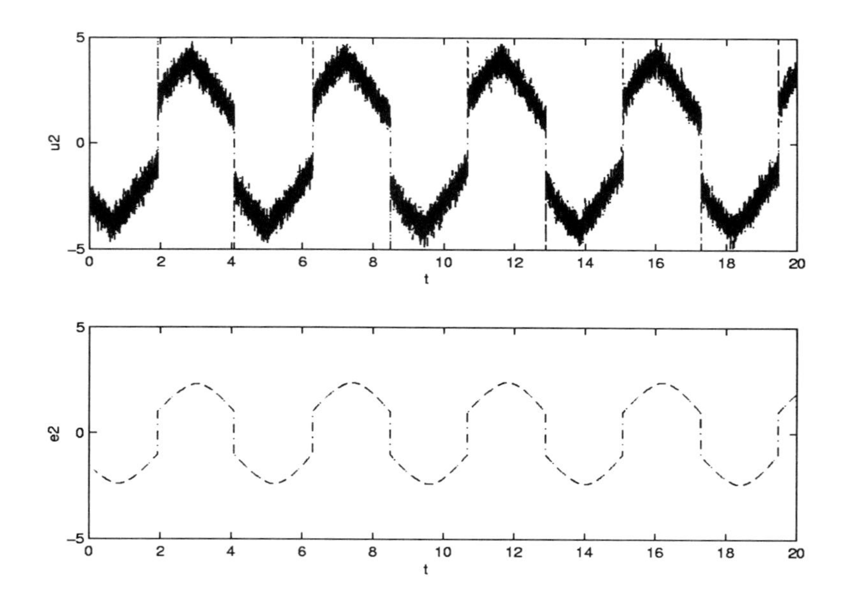

Fig. 4.18. On-line identification experiment: Unknown structure only and noisy measurements

Example 4.13: Both PID structure and parameters are unknown

Consider the process:

$$G_p(s) = \frac{1.2e^{-10s}}{(5s+1)(2.5s+1)}, \tag{4.82}$$

and assume the following PID controller has been commissioned and in force

$$G_c(s) = 0.771\left(1 + \frac{1}{16.31s} + 4.077s\right), \tag{4.83}$$

Both PID structure and parameters are assumed unknown in this example. The first phase of the experiment is carried out with the D part of the controller turned off. The PI controller is identified according to (4.45),(4.46) as

$$G_c(s) = 0.771\left(1 + \frac{1}{15.45s}\right), \tag{4.84}$$

The signals concerned are shown in Fig. 4.19.

In the second phase, the D part is reconnected and the experiment repeated. The signals are shown in Fig. 4.20. From (4.45), $T_d = 3.9$. With all three parameters, the PID controller is constructed in both parallel and series form, and the corresponding performance index J computed. For the parallel structure, $J = 0.0101$ and for the series structure, $J = 0.1330$. The parallel structure is correctly identified.

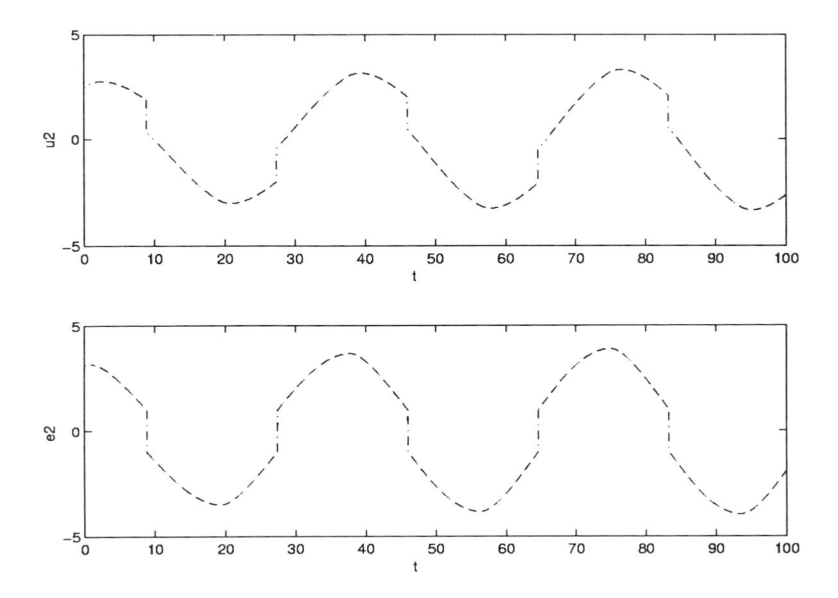

Fig. 4.19. On-line identification experiment - Unknown PID structure and parameters; T_d disabled

Example 4.14: PID assessment and direct autotuning

In this example, it will be demonstrated how to assess and tune (direct method) the controller within one single on-line experiment when only r and y are accessible.

Consider the following process:

$$G_p(s) = \frac{1}{(s+1)^2} e^{-s}. \tag{4.85}$$

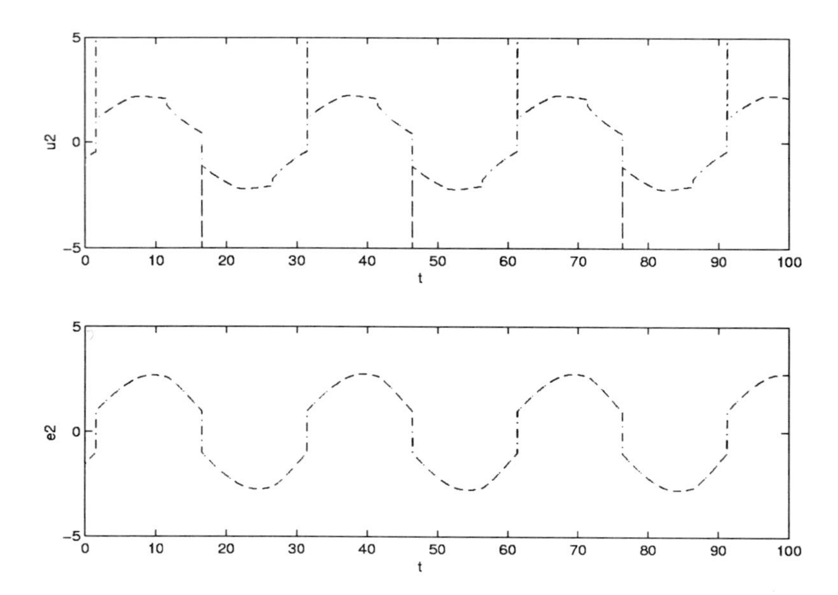

Fig. 4.20. On-line identification experiment - Unknown PID structure and parameters; T_d enabled

The parallel type PID controller used is,

$$G_c(s) = 2\left(1 + \frac{1}{3.559s} + 0.55s\right). \tag{4.86}$$

From the first closed-loop step response, it is observed that this controller is not well tuned. The stablity performance is poor; it takes several oscillations for system to reach the steady state. The on-line relay experiment begins at $t = 40$, the assessment factors are now calculated from the limit cycles as $\beta = 2.5$ and $\gamma = 0.1$ ($\tilde{G}(j\omega) = -0.12 + 0j$). The controller is retuned and the parameters calculated as,

$$G_c(s) = 0.62\left(1 + \frac{1}{1.1s} + 0.275s\right), \tag{4.87}$$

The more stable subsequent performance is shown in Fig. 4.21. The new assessment factors are calculated as $\beta = 1$ and $\gamma = 0$.

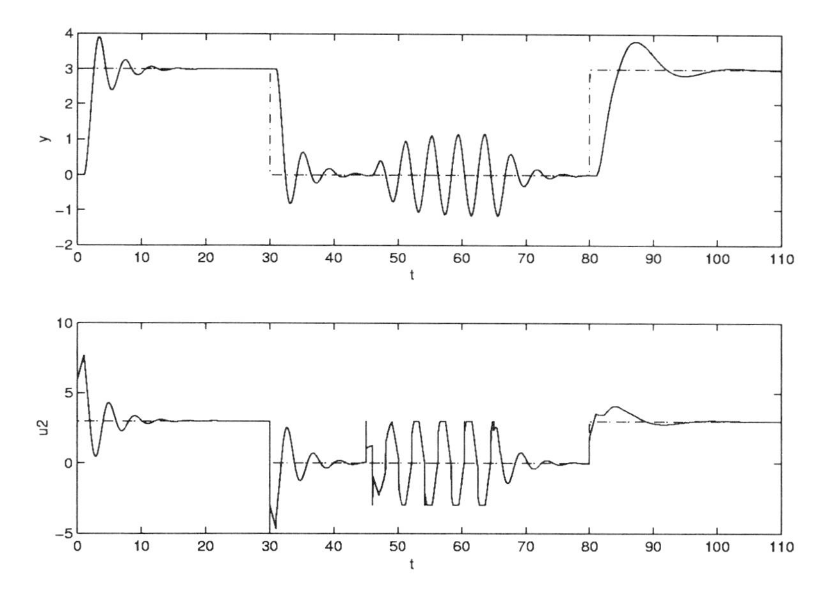

Fig. 4.21. PID controller assessment and auto-tuning (Direct Method)

Example 4.15: PID assessment and indirect autotuning

Consider the same process and original controller as previous example, the indirect method is applied which yields the process model as

$$\tilde{G}_p(s) = \frac{1}{1 + 1.778s} e^{-1.5007s}. \tag{4.88}$$

The optimal PI tuning method via the LQR approach is applied based on this model. With the controller accordingly commissioned, a setpoint change occurs at $t = 80$. The good subsequent control performance is shown in Fig. 4.22. The new assessment factors are calculated as $\beta = 0.067$ and $\gamma = 0.32$.

Real-time Experiment. The on-line PID auto-tuning method is applied in a real-time experiment to *KI-100 Dual Process Simulator* from Kent Ridge Instruments. This is an analog process simulator which can be configured to simulate a wide range of industrial processes with different kinds of dynamics and at different levels of noise. The simulator is connected to a PC via an integrated ADDA and DSP control board *dSPACE DS1102. Simulink and RTW Toolbox* from *Mathworks* is the control development platform. The sampling period used in the software is 0.001 sec. The process used for real time experiment is configured to:

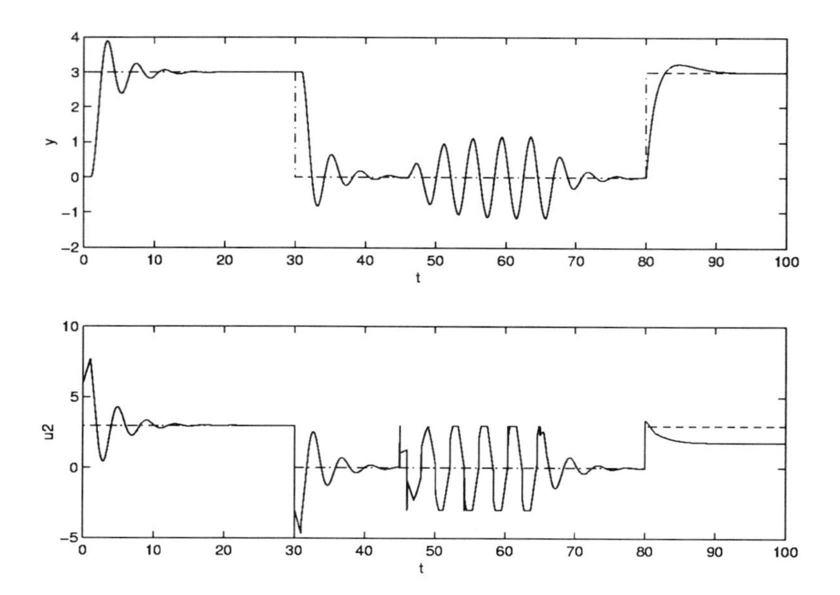

Fig. 4.22. PID controller assessment and auto-tuning(Indirect Method)

$$G_p(s) = \frac{1}{(2s+1)^2} e^{-1.00s}. \tag{4.89}$$

From the first closed-loop step response, it is observed that this controller is not well tuned. The stablity performance is poor; it takes several oscillations for system to reach the steady state. The on-line relay experiment begins at $t = 160s$, the assessment factors are now calculated from the limit cycles as $\beta = 2.5$ and $\gamma = 0.1$ ($\tilde{G}(j\omega) = -0.2 + 0j$). The structure and parameters of the PID controller are correctly identified as follows:

$$G_c(s) = 1.536 \left(1 + \frac{1}{2.06s} + 0.52s \right). \tag{4.90}$$

The controller is retuned and the more optimal set of control parameters calculated as,

$$G_c(s) = 0.768 \left(1 + \frac{1}{2.074s} + 0.519s \right). \tag{4.91}$$

The improved subsequent performance is shown in Fig. 4.23.

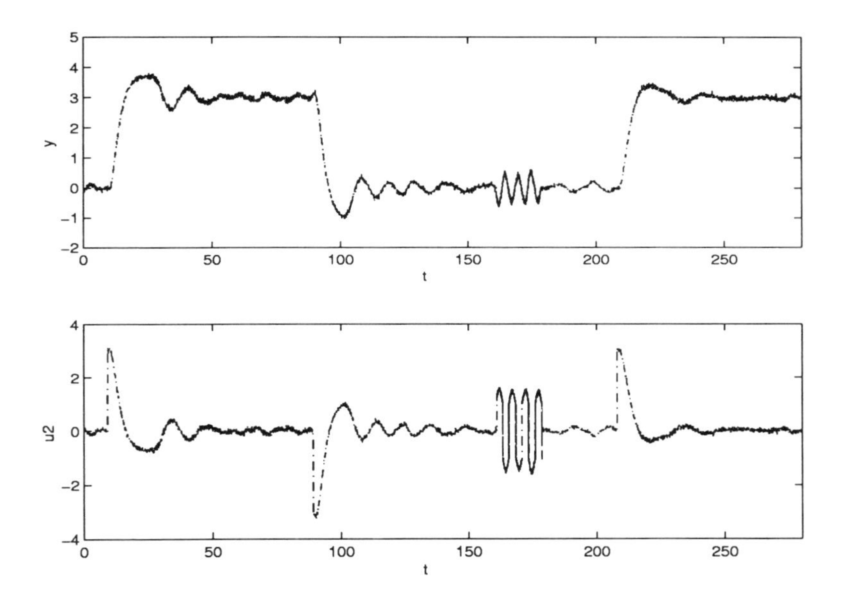

Fig. 4.23. Real-time experiment

4.5 FFT on Relay Transients

It is noted that the basic relay method and the modifications mentioned in the previous sections employ only stationary relay oscillations with the relay transients in process input and output totally ignored. The advantages are that the resultant estimations are simple and robust to noise, non-zero initial condition and disturbance. On the other hand, however, the nature of such methods greatly limits identifiability of process dynamics since dynamic information is mainly contained in process transient response to a test. In this section, a method for identifying multiple points on process frequency response from a single relay test is presented with appropriate processing of the relay transient. Since the input and output transients resulting from a relay test cannot be directly transferred to frequency response meaningfully using FFT, a decay exponential is introduced to the process input and output so that the modified input and output will approximately decay to zero in a finite time interval. FFT is then employed to obtain the process frequency response. This method uses the same relay test as in the basic method and yet it can yield as many frequency response points as desired, and the estimation result is accurate.

4.5.1 The FFT-Relay Method

Consider again the relay feedback system shown in Fig. 4.3. If the process has a phase lag of at least π radians, the relay feedback will cause the system to oscillate. Assume that the system is at a steady-state before a relay feedback is applied at $t = 0$. The steady-state condition is an ordinary assumption which can also be found in other identification schemes such as a step test. For example, the Honeywell TDC $3000^{\times}$ Robust PID Controller design requires a steady state at the beginning of a step identification test (Honeywell, 1995). In a relay test, the process output $y(t)$ and input $u(t)$ are recorded from the time the test starts until the system reaches a stationary oscillation.

In order to motivate the development, it is imperative to note why a direct application of the Fast Fourier Transform (FFT) to relay transient fails to obtain process frequency response. If the process is described by $Y(s) = G_p(s)U(s)$, it follows that $G_p(j\omega) = Y(j\omega)/U(j\omega)$. The question is how to obtain $Y(j\omega)$ and $U((j\omega)$ from $y(t)$ and $u(t)$ numerically. At the first glance, it seems that $Y(j\omega)$ and $U(j\omega)$ can be easily obtained by taking FFT of $y(t)$ and $u(t)$ directly. Unfortunately, this operation is false and meaningless, as $y(t)$ and $u(t)$ are neither absolutely integrable nor strictly periodic. To see the mechanism of this failure more clearly, $y(t)$ or $u(t)$ is decomposed into the transient parts Δy or Δu and the periodic stationary cycle parts y_s or u_s as:

$$y(t) = \Delta y + y_s, \tag{4.92}$$

and

$$u(t) = \Delta u + u_s. \tag{4.93}$$

Once $y(t)$ and $u(t)$ reach the stationary oscillation status at $t = T_f$, both Δy and Δu are approximately zero afterwards. The Fourier transform of Δy is then

$$\begin{aligned} \Delta Y(j\omega) &= \int_0^{\infty} \Delta y(t) e^{-j\omega t} dt \\ &\approx \int_0^{T_f} \Delta y(t) e^{-j\omega t} dt. \end{aligned} \tag{4.94}$$

(4.94) can be computed at discrete frequencies with the standard FFT technique. FFT of one period of the periodic parts y_s (u_s) will actually give the scaled coefficients of the corresponding Fourier series of y_s (u_s) or the scaled amplitudes of the impulses of the extended Fourier transform of y_s or u_s (Cartwright, 1990). The FFT results from the transient parts and the periodic parts thus have different meanings and they cannot be added together. This means that the FFT cannot directly be applied to $y(t)$ or $u(t)$ to obtain the process frequency response.

To overcome the above obstacle, a decay exponential $e^{-\alpha t}$ ($\alpha > 0$) is here introduced to form

$$\tilde{y}(t) = y(t) e^{-\alpha t}, \tag{4.95}$$

and

$$\tilde{u}(t) = u(t) e^{-\alpha t}. \tag{4.96}$$

so that $\tilde{u}(t)$ and $\tilde{y}(t)$ decay to zero exponentially as t approaches infinity. Applying the Fourier Transform to (4.95) and (4.96) yields

$$\tilde{Y}(j\omega) = \int_0^\infty \tilde{y}(t)e^{-j\omega t}dt = \int_0^\infty y(t)e^{-\alpha t}e^{-j\omega t}dt = Y(j\omega + \alpha), \qquad (4.97)$$

and

$$\tilde{U}(j\omega) = \int_0^\infty \tilde{u}(t)e^{-j\omega t}dt = \int_0^\infty u(t)e^{-\alpha t}e^{-j\omega t}dt = U(j\omega + \alpha), \qquad (4.98)$$

The integral intervals in (4.97) and (4.98) are infinite and digital computation of the infinite interval integration is non-trivial. However, due to the introduction of the decay exponential $e^{-\alpha t}$, $\tilde{y}(t)$ and $\tilde{u}(t)$ are approximately zero after a certain time span. The infinite interval integration problem actually becomes a finite integration one. Thus, $\tilde{Y}(j\omega)$ can be computed at discrete frequencies with the standard FFT technique (Ramirez 1985). Suppose that $y(kT_s)$, $k = 0, 1, 2, \cdots, N-1$, are samples of $y(t)$, where T_s is the sampling time interval. N is chosen such that $y((N-1)T_s)$ has reached a stationary oscillation and the decay coefficient α is selected such that $\tilde{y}((N-1)T_s)$ formed from (4.95) has approximately decayed to zero. It then follows that

$$\begin{aligned}\tilde{Y}(j\omega_i) &\approx T_s \sum_{k=0}^{\infty} \tilde{y}(kT_s)e^{-j\omega_i kT_s} \approx T_s \sum_{k=0}^{N-1} \tilde{y}(kT_s)e^{-j\omega_i kT_s} \\ &= FFT(\tilde{y}(kT_s)), \qquad i = 1, 2, \cdots, M,\end{aligned} \qquad (4.99)$$

where $M = \frac{N}{2}$ and $\omega_i = 2\pi i/(NT_s)$. $\tilde{U}(j\omega)$ can be similarly computed by taking FFT of $\tilde{u}(kT_s)$. Therefore, the shifted process frequency response $G_p(j\omega_i + \alpha)$ is given by:

$$G_p(j\omega + \alpha) = \frac{Y(j\omega_i + \alpha)}{U(j\omega_i + \alpha)} = \frac{\tilde{Y}(j\omega_i)}{\tilde{U}(j\omega_i)}, \qquad i = 1, 2, \cdots, M. \qquad (4.100)$$

Thus, the shifted process frequency response $G_p(j\omega + \alpha)$ is obtained. This identified shifted process frequency response is usually enough for process modeling and controller design (Wang et al., 1997). However, if $G_p(j\omega)$ other than $G_p(j\omega + \alpha)$ is needed, the inverse FFT of $G_p(j\omega + \alpha)$ can be taken as

$$\tilde{g}(kT_s) := FFT^{-1}(G_p(j\omega_i + \alpha)) = G_p(kT_s)e^{-\alpha kT_s}. \qquad (4.101)$$

It then follows from $\tilde{g}(kT_s)$ that the process impulse response $G_p(kT_s)$ is

$$G_p(kT_s) = \tilde{g}(kT_s)e^{\alpha kT_s}. \tag{4.102}$$

Applying the FFT again to $G_p(kT_s)$ would result in the process frequency response in $G_p(j\omega_i)$ form:

$$G_p(j\omega_i) = FFT(G_p(kT_s)). \tag{4.103}$$

Unlike the method of Li et al. (1991), one test is sufficient for the multiple-point frequency response identification and little prior knowledge of process is required with this method.

Parameter Selection. In order to implement this identification procedure in real time, values must be given for the decay factor α and the time span $T_f = (N-1)T_s$ required in the FFT computations (4.99). Consider first the selection of T_f first. Recall that the method can produce as many frequency response points as desired. Suppose that the number of the frequency response points to be identified from zero frequency to phase-crossover frequency ω_π is M. M should be specified by the system designers and will actually depend on the method which is used to tune the controller. It follows from (4.99) that the M frequency response points recovered by the FFT method are at the discrete frequencies $0, \Delta\omega, 2\Delta\omega, \cdots, (M-1)\Delta\omega$, where $\Delta\omega = \omega_{i+1} - \omega_i = 2\pi/NT_s$, is the frequency increment. The definition of M means that $\omega_\pi \approx (M-1)\Delta\omega$, thus

$$\omega_\pi \approx (M-1)\frac{2\pi}{NT_s}. \tag{4.104}$$

On the other hand, one can measure the oscillation period T_π on line when a relay test is performed and ω_π can be estimated (Astrom, 1988) as

$$\omega_\pi \approx \frac{2\pi}{T_\pi}. \tag{4.105}$$

Equations (4.104) and (4.105) yield

$$N \approx (M-1)\frac{T_\pi}{T_s}, \tag{4.106}$$

where M should be specified a priori and should be large enough to ensure that the stationary oscillation is reached. The corresponding time span is then $T_f = (N-1)T_s$.

For the decay coefficient α, its value should be such that $\tilde{y}(t)$ and $\tilde{u}(t)$ decay nearly to zero when the time approaches T_f, regardless of non-zero $y(t)$ and $u(t)$. It is this decay coefficient α that enables the infinite integral of Fourier transform to be approximately replaced by the finite digital integral in FFT. To have a good approximation in (4.99), the decay coefficient α should

$$e^{-\alpha T_f} \leq \varepsilon, \tag{4.107}$$

or

$$\alpha \geq -\frac{\ln \varepsilon}{T_f}, \tag{4.108}$$

where ε is the specified threshold and usually takes the value of $10^{-4} \sim 10^{-6}$.

Noise Issue. Noise is a significant issue in system identification. It is apparent (Ljung, 1987) that in almost all identification methods a low noise-to-signal ratio is required. As for the measurement noise in the relay test, Astrom et al. (1984) pointed out that a hysteresis in the relay is a simple way to reduce the influence of the measurement noise. The width of hysteresis should be bigger than the noise band (Astrom et al., 1988) and is usually chosen as two times larger than the noise band (Hang et al., 1993b). Filtering is another possibility (Astrom et al., 1984). The measurement noise is usually of high frequency while the process frequency response of interest for control analysis and design is usually in the low frequency region. It is found from experiments that a low pass filter can be employed to reduce the measurement noise effect further if the estimation of $\tilde{G}(j\omega)$ is needed. But the estimation of $\tilde{G}(j\omega+\alpha)$ is not sensitive to noise and no filter is needed if it stops with $\tilde{G}(j\omega+\alpha)$. In particular, the process frequency response in $[0, \omega_\pi]$ is mostly critical for controller design. It is found in the experiments the measurement noise is indeed in the fairly high frequency region. Therefore, the cut-off frequency of the filter is determined with respect to the process frequency region of interest. The FFT-Relay method can reject noise quite effectively with the above anti-noise measures.

4.5.2 Simulation study

The FFT-Relay frequency response identification method has been applied to several typical processes. For assessment of accuracy, the identification error is here measured by worst case error

$$ERR = \max_i \left\{ \left\| \frac{\tilde{G}(j\omega_i) - G_p(j\omega_i)}{G_p(j\omega_i)} \right\| \times 100\%, i = 1, 2, \ldots, M \right\}, \quad (4.109)$$

where $G_p(j\omega_i)$ and $\tilde{G}(j\omega_i)$ are the actual and the estimated process frequency responses respectively. The Nyquist curve for phase ranging from 0 to $-\pi$ is being considered since this part is most significant for control design.

Example 4.16:

This example is adopted from Li et al. (1991) as:

$$G_p(s) = \frac{e^{-2s}}{10s + 1}.$$

The model estimated by Li et al. (1991) is $\tilde{G}(s) = \frac{0.988e^{-2s}}{8.02s+1}$ and its identification error is $ERR = 22.32\%$. In Li's method, the dead-time is assumed to be known. For the FFT-Relay method, a relay feedback is applied to the process. The process output and input are logged. $y(t)$ and $\tilde{y}(t)$ of the relay test are shown in 4.24(a) while $u(t)$ and $\tilde{u}(t)$ are exhibited in Fig. 4.24(b). The frequency response identified by the FFT-Relay method using MATLAB is shown in Fig. 4.25. The ERR is 0.26%. This indicates that the FFT-Relay method provides a much more accurate process frequency response.

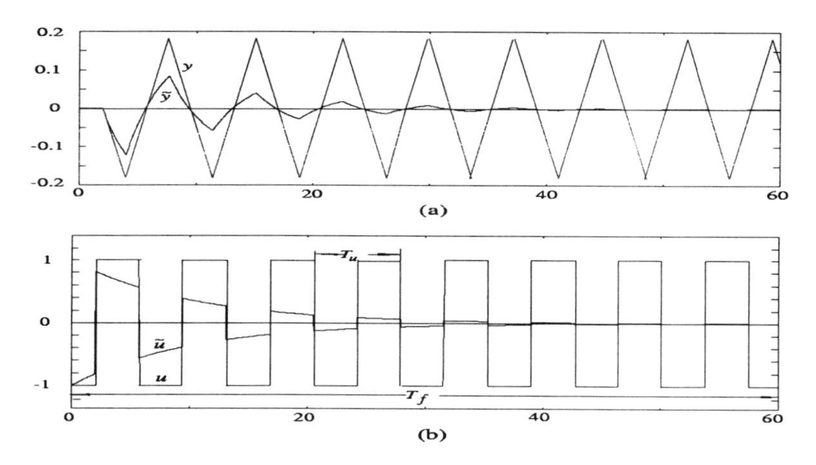

Fig. 4.24. Signals under relay feedback

The method was tested on the *Dual Process Simulator KI 100.* In the context of system identification, noise-to-signal ratio (Haykin, 1989) is usually defined as:

$$N_1 = \frac{mean\ power\ spectrum\ density\ of\ noise}{mean\ power\ spectrum\ density\ of\ signal},$$

that is the noise-to-signal power spectrum ratio or

$$N_2 = \frac{mean(abs(noise))}{mean(abs(signal))},$$

that is noise-signal mean ratio. A few examples of real-time testing are presented as follows.

Example 4.17:

Consider a first order plus dead-time process

$$G_p(s) = \frac{1}{5s+1}e^{-5s}.$$

The low pass filter is selected as a Butterworth low pass filter whose cut-off frequency is chosen as $3 \sim 5\omega_\pi$. Other anti-noise actions discussed in Section

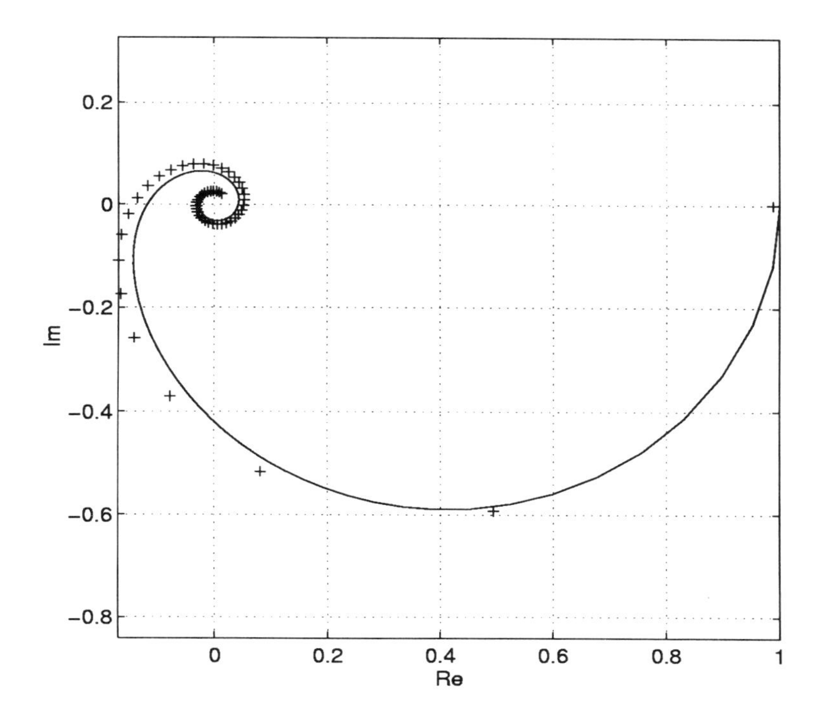

Fig. 4.25. Nyquist plot of $G_p(j\omega)$. $\cdots$ actual, + estimated

4.5.1 are also taken. Without additional noise, the noise-to-signal ratio N_1 of the inherent noise in the test environment is 0.06% ($N_2 = 2\%$). The identification error ERR in terms of the shifted frequency response $G_p(j\omega + \alpha)$ is 2.77% with $\alpha = 0.04$.

To observe the noise effects, extra noise is introduced with the noise source in the *Simulator*. Time sequences of $y(t)$ and $u(t)$ in a relay test under $N_1 = 10\%$ ($N_2 = 35\%$) are shown in Fig. 4.26. The first part of the test in Fig. 4.26 ($t = 0 \sim 30$) is the "listening period", in which the noise bands of $y(t)$ and $u(t)$ at steady state are measured. The estimated frequency response points are shown in Fig. 4.27(a).

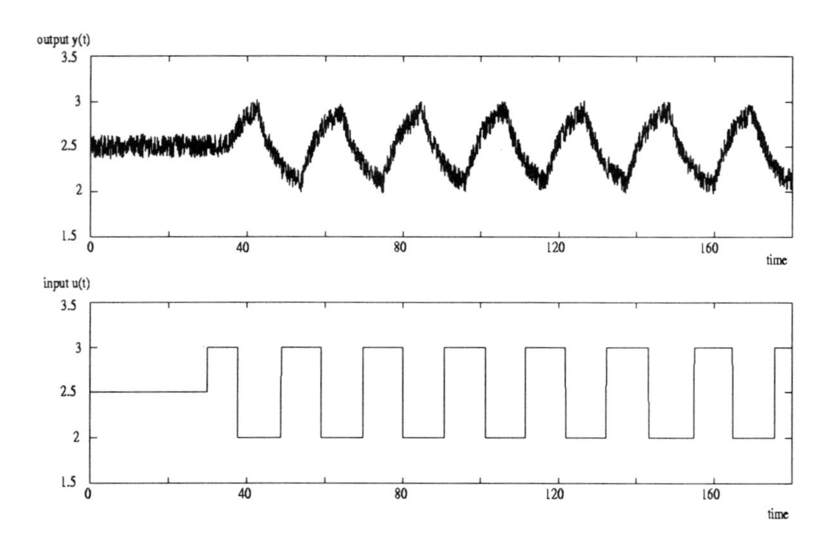

Fig. 4.26. Real-time signals under relay feedback

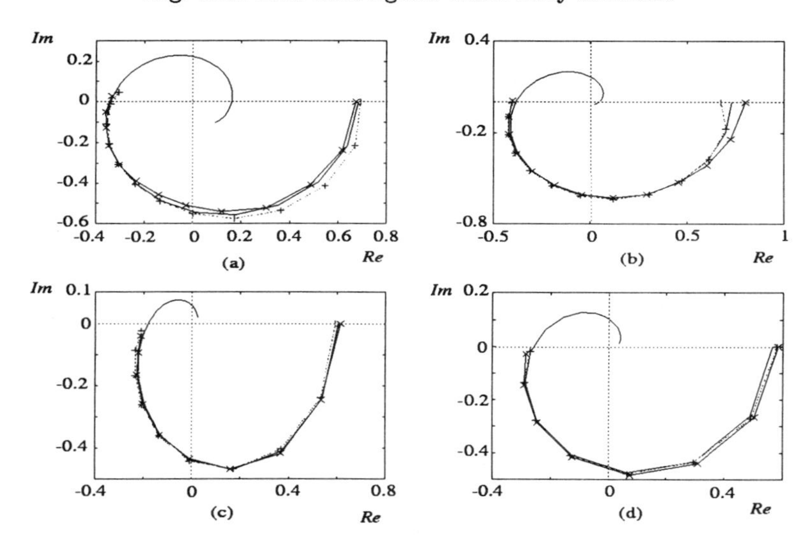

Fig. 4.27. Nyquist Plot of $G_p(j\omega+\alpha)$.$-$ actual,$\cdots+\cdots$ estimated($N_1 = 1\%$),$\cdots\times\cdots$ estimated($N_1 = 10\%$).

Example 4.18:

For a multi-lag high order process:

$$G_p(s) = \frac{1}{(s+1)^8},$$

the actual and the estimated frequency responses are plotted in Fig. 4.27(b).

Example 4.19:

Consider a process which has different poles

$$G_p(s) = \frac{1}{(s+1)(5s+1)^2} e^{-2.5s}.$$

The actual and the estimated frequency responses are presented in Fig. 4.27(c). The accuracy of the estimated frequency response is excellent.

Example 4.20: For a non-minimum phase plus dead-time process

$$G_p(s) = \frac{1-s}{(2s+1)^4(5s+1)} e^{-s},$$

the actual and the estimated Nyquist curves are shown in Fig. 4.27(d). The accuracy of the FFT-Relay method is evident.

Table 4.3 shows the identification accuracy of the above four real-time examples under different noise levels. It can be seen from simulation that multiple frequency response points have been identified simultaneously with one single relay experiment and this saves testing time greatly. The method is accurate, especially in the critical frequency range $[0, \omega_\pi]$, provided that the process is linear and no disturbance exists.

4.6 Frequency Response - Transfer Function Conversion

The identification methods described in the previous sections may yield the frequency response of the process at certain frequencies. However, for simulation or control purposes, transfer function models fitted to the frequency response estimation are more directly useful. Pintelon et al. (1994) give a good survey on parameter identification of transfer functions without dead-time in frequency domain. With dead-time, the parameter identification is usually a nonlinear problem (Palmor, 1994) and difficult to solve. In this section, the identification of common low-order transfer function models from the frequency response estimation will be illustrated. Such models have proven

Table 4.3. Identification error (ERR)

Process	Noise Levels		
	$N_1 = 1\%$	6%	10%
	$N_2 = 13\%$	25%	35%
$\frac{1}{5s+1}e^{-5s}$	6.14%	7.96%	9.67%
$\frac{1}{(s+1)^8}$	7.42%	8.97%	9.85%
$\frac{1}{(s+1)(5s+1)^2}e^{-2.5s}$	3.75%	6.52%	8.65%
$\frac{1-s}{(2s+1)^4(5s+1)}e^{-s}$	2.93%	4.48%	8.69%

to be relevant and adequate particularly in the process control industry. Algorithms are developed to calculate the parameters of these approximate transfer function models from the frequency response information acquired using the techniques discussed in the earlier sections. Such an identification technique is readily automated, and as such, it is useful for auto-tuning applications.

4.6.1 Single and multiple lag processes

It is well known that many processes in the industry are of the low-order dynamics, and they can be adequately modeled by a rational transfer function of the following form:

$$\tilde{G}_p(s) = \frac{K_p}{Ts+1}e^{-Ls}, \tag{4.110}$$

where K, T and L are real parameters to be estimated. It is desirable to distinguish this class of processes from higher-order ones since controller design and other supervisory tasks are often much simplified with the use of the model.

Two points on the process Nyquist curve are sufficient to determine the three parameters of (4.110). Assume that the following Nyquist points of the process $G_p(s)$ have been obtained: $G_p(j\omega_1)$ and $G_p(0)$, $\omega_1 \neq 0$. The process static gain is given by

$$K_p = G_p(0). \tag{4.111}$$

Equating the process and model at $\omega = \omega_1$ yields

$$|G_p(j\omega_1)| := \frac{1}{k_1} = \frac{K}{\sqrt{1+T^2\omega_1^2}} \ , \tag{4.112}$$

$$\arg\, G_p(j\omega_1) := \phi_1 = -\arctan T\omega_1 - L\omega_1. \tag{4.113}$$

It follows that

$$T = \frac{1}{\omega_1}\sqrt{k_1^2K^2 - 1} \ , \tag{4.114}$$

$$L = -\frac{1}{\omega_1}(\phi_1 + \arctan\omega_1 T). \tag{4.115}$$

A more general model adequate for processes with a monotone open-loop step response is of the following form:

$$\tilde{G}_p(s) = \frac{K_p}{(Ts+1)^n}e^{-Ls}, \tag{4.116}$$

where K_p, T and L are real parameters to be estimated, and $n \in Z^+$ is an order estimate of the process.

One method for order estimation based on this model was proposed by Lundh (1991). He noted that the maximal slope of the frequency response magnitude is a measure of the process complexity, and he uses the slope in the vicinity of the critical frequency to estimate the relative degree of the process, choosing one of three possible models to represent the process. In his method, FFTs are performed on the input and output of the process, from which the amplitudes of the first and third harmonics of the frequency spectrum are used to compute the amplitude gains at these two frequencies. From these frequencies, the slope of the frequency response magnitude at the geometrical mean of the harmonics is then calculated. A simpler method for order estimation is presented here which does not require a Fourier transform on the input and output signals of the process.

Assume that the following Nyquist points of the process $G_ps)$ have been obtained: $G_p(j\omega_1)$ and $G_p(j\omega_2)$, $\omega_1 \neq \omega_2$ and $\omega_1, \omega_2 \neq 0$. Equating the gain of the process and model at ω_1 and ω_2, one has

$$|G_p(j\omega_1)| := \frac{1}{k_1} = \frac{K_p}{\left(\sqrt{1+T^2\omega_1^2}\right)^n}, \tag{4.117}$$

$$|G_p(j\omega_2)| := \frac{1}{k_2} = \frac{K_p}{\left(\sqrt{1+T^2\omega_2^2}\right)^n}. \tag{4.118}$$

Simplifying (4.117) and (4.118), it follows that

$$\left[1+\left(\frac{\omega_2}{\omega_1}\right)^2\left((K_pk_1)^{2/n}-1\right)\right]^{n/2} = K_pk_2. \tag{4.119}$$

Equating the phase of the process and model at ω_1 and ω_2, it follows

$$\arg G_p(j\omega_1) := -\phi_1 = -n\arctan T\omega_1 - L\omega_1, \tag{4.120}$$

$$\arg G_p(j\omega_2) := -\phi_2 = -n\arctan T\omega_2 - L\omega_2 \tag{4.121}$$

A simple algorithm to obtain the parameters in (4.116) is outlined below. $\tilde{n}$ is a specified upper bound on the order of the process.

- Iterate from $n = 1$ to $n = \tilde{n}$.
  - Compute K_p from (4.119), T from (4.117) or (4.118), and L from (4.120).
  - Compute the cost function derived from (4.120),

    $$J(n) = |-\phi_2 + n\arctan T\omega_2 + L\omega_2|$$

    with the values of n, K_p, T and L above.
- At the end of the iteration, a suitable set of model parameters (n, K_p, T, L) corresponds to $n = n_{\min}$, where

  $$J(n_{\min}) = \min_n \{J(n)\}.$$

4.6.2 Second-order modeling

In this subsection, a non-iterative method is presented for the following stable low-order plus dead-time model:

$$\tilde{G}_p(s) = \frac{1}{as^2 + bs + c} e^{-Ls}, \tag{4.122}$$

which can represent both monotonic and oscillatory processes.

Transfer Function Modeling from $G_p(j\omega)$. Assume that the process frequency response $G_p(j\omega_i)$, $i = 1, 2, \cdots, M$, is available, and it is required to be fitted into $\tilde{G}(s)$ in (4.122) such that

$$G_p(j\omega_i) = \frac{1}{(j\omega_i)^2 a + j\omega_i b + c} e^{-j\omega_i L}, \quad i = 1, 2, \cdots, M. \tag{4.123}$$

The determination of the parameters a, b, c and L in (4.123) seems to be a nonlinear problem. One way of solving this problem is to find the optimal a, b and c given L and iteratively determine L by some searching algorithm. To avoid the iteration, the magnitude of both sides of (4.123) is taken as

$$\left[\omega_i^4 \; \omega_i^2 \; 1\right] \theta = \frac{1}{|G_p(j\omega_i)|^2}, \quad i = 1, 2, \cdots, M, \tag{4.124}$$

where $\theta = \left[a^2 \; b^2 - 2ac \; c^2\right]^T$. This action shields the effects from L and forms a system of linear equations (4.124) in θ which can be solved for θ with the linear least squares method. Then, $\left[a \; b \; c\right] = \left[\sqrt{\theta_1} \; \sqrt{\theta_2 + 2\sqrt{\theta_1 \theta_3}} \; \sqrt{\theta_3}\right]$. In addition, the phase relation in (4.123) gives

$$\omega_i L = -arg[G_p(j\omega_i)] - tan^{-1}\left(\frac{b\omega_i}{c - a\omega_i^2}\right), \quad i = 1, 2, \cdots, M. \tag{4.125}$$

Obviously, L can be obtained again with the least squares method.

Transfer Function Modeling from $G_p(j\omega + \alpha)$. With the method in Section 4.5, $G_p(j\omega + \alpha)$ is a more direct product than $G_p(j\omega)$. Thus, direct transfer function modeling from $G_p(j\omega + \alpha)$ is of more interest in this case. Determining the parameters a, b, c and L from $G_p(j\omega + \alpha)$ is not an easy job. Here, a possible solution is presented.

When the process frequency responses $G_p(j\omega + \alpha)$, $i = 1, 2, \cdots, M$ are available, they are fitted into $\tilde{G}(s)$ in (4.122) such that

$$G_p(j\omega_i+\alpha)=\frac{1}{(j\omega_i+\alpha)^2a+(j\omega_i+\alpha)b+c}e^{-(j\omega_i+\alpha)L},\quad i=1,2,\cdots,M. \tag{4.126}$$

Taking the magnitude of both sides of (4.126) yields

$$|G_p(j\omega_i+\alpha)|=\frac{1}{\sqrt{(\nu-a\omega_i^2)^2+\omega_i^2(b+2a\alpha)^2}}e^{-\alpha L},\ i=1,2,\cdots,M. \tag{4.127}$$

where $\nu=a\alpha^2+b\alpha+c$. $\tilde{a},\tilde{b}$ and $\tilde{\nu}$ are computed from

$$|G_p(j\omega_i+\alpha)|=\frac{1}{\sqrt{(\tilde{\nu}-\tilde{a}\omega_i^2)^2+\omega_i^2(\tilde{b}+2\tilde{a}\alpha)^2}},\ i=1,2,\cdots,M. \tag{4.128}$$

(4.128) can be changed into is a system of linear equations in $\tilde{\theta}$ as

$$\begin{bmatrix}\omega_i^4 & \omega_i^2 & 1\end{bmatrix}\tilde{\theta}=\frac{1}{|G_p(j\omega_i+\alpha)|^2},\quad i=1,2,\cdots,M, \tag{4.129}$$

where $\tilde{\theta}=\begin{bmatrix}\tilde{a}^2 & \tilde{b}^2+4\tilde{a}\tilde{b}\alpha+4\tilde{a}^2\alpha^2-2\tilde{a}\tilde{\nu} & \tilde{\nu}^2\end{bmatrix}^T$. (4.129) can be solved with the linear least squares method and $\tilde{a},\tilde{b}$ and $\tilde{\nu}$ are obtained. Equations (4.127) and (4.128) yield

$$\frac{1}{\sqrt{(\nu-a\omega_i^2)^2+\omega_i^2(b+2a\alpha)^2}}e^{-\alpha L}=\frac{1}{\sqrt{(\tilde{\nu}-\tilde{a}\omega_i^2)^2+\omega_i^2(\tilde{b}+2\tilde{a}\alpha)^2}} \tag{4.130}$$

for every ω_i, $i=1,2,\cdots,M$. Matching the coefficients of like power of ω_i in (4.130) yields

$$\begin{cases}a=e^{-L\alpha}\tilde{a}\\ b=e^{-L\alpha}\tilde{b}\\ c=e^{-L\alpha}\tilde{c}\end{cases} \tag{4.131}$$

for stable process and $\alpha>0$. Phase relation of (4.126) leads to

$$\omega_i L = -\arg[G_p(j\omega_i + \alpha)] - tan^{-1}(\frac{b\omega_i + 2a\alpha\omega_i}{\nu - a\omega_i^2}), \ i = 1, 2, \cdots, M, \tag{4.132}$$

Using (4.131), (4.132) becomes

$$\omega_i L = -\arg[G_p(j\omega_i + \alpha)] - tan^{-1}(\frac{\tilde{b}\omega_i + 2\tilde{a}\alpha\omega_i}{\tilde{\nu} - \tilde{a}\omega_i^2}), \ i = 1, 2, \cdots, M, \tag{4.133}$$

where $\tilde{a}, \tilde{b}$ and $\tilde{\nu}$ have been obtained by (4.129). L then can be obtained with the least squares method. Once L is known, a, b and c in (4.126) can then be determined by (4.131).

Table 4.4. Transfer function modeling

Process	Process Models	
	from $G_p(j\omega)$	from $G_p(j\omega + \alpha)$
$\frac{1}{10s+1}e^{-2s}$	$\frac{1}{(9.99s+1)}e^{-2.01s}$	$\frac{1}{(10.00s+1)}e^{-2.00s}$
$\frac{1}{(s+1)^{10}}$	$\frac{1}{8.29s^2+5.06s+1}e^{-5.02s}$	$\frac{1}{8.16s^2+5.05s+1}e^{-5.02s}$
$\frac{-s+1}{(s+1)^5}e^{-2s}$	$\frac{1}{2.68s^2+3.04s+0.999}e^{-4.90s}$	$\frac{1}{2.63s^2+3.04s+1.01}e^{-4.91s}$
$\frac{1}{s^2+s+1}e^{-s}$	$\frac{1}{s^2+s+1}e^{-1.00s}$	$\frac{1}{1.0s^2+1.0s+1.0}e^{-1.00s}$

Simulation is carried out on a number of processes to illustrate the transfer function modeling procedures. The resultant transfer functions derived from the estimated $G_p(j\omega)$ and $G_p(j\omega + \alpha)$ are listed in Table 4.4, which shows that the two methods produce very similar results. The actual and estimated Nyquist plots are shown in Fig. 4.28(a)-Fig. 4.28(d) and they fit each other very well.

4.7 Continuous Self-Tuning of PID Control

In many industrial processes, dynamics may drift significantly from time to time. If no corrective action is taken, the system performance naturally deteriorates. Thus, the controller being used should adapt itself to such dynamics

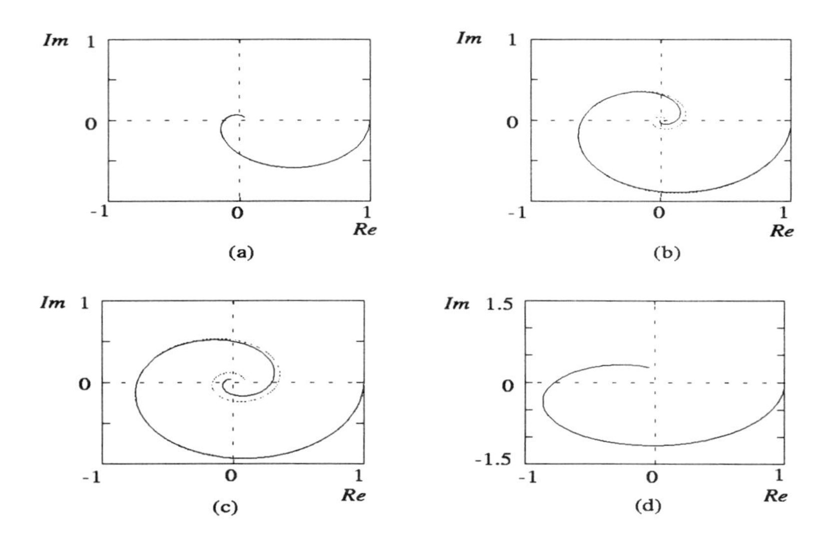

Fig. 4.28. Nyquist Plot of $G_p(j\omega+\alpha)$.– actual,$\cdots+\cdots$ estimated($N_1 = 1\%$),$\cdots\times\cdots$ estimated($N_1 = 10\%$).

change in order to ensure performance specifications. Continuous self-tuning is a powerful technique to realize this objective. However, a self-tuner needs to be switched off during a load change (Hang et al., 1993 and 1995). Experience has also shown that many self-tuning PID controllers are not being operated in the continuous self-tuning mode (Hang et al., 1993). In fact, when the process perturbs slowly, the continuous adaptive controller will not be necessary. Frequent controller adjustments will be enough. Some rule-based adaptation schemes (Hang et al., 1993; Astrom and Hagglund, 1995) are of this type. They make use of the process response under a set-point change or a load disturbance to re-tune their controllers. Based on the process response patterns, the controllers will be modified using the pre-setting rules. These rules are generally quite complex in order to work efficiently in different situations and are not easy to devise. The rules are usually empirical and the resultant performance mainly depends on the rule-formulator experience. These methods are more suitable for those controllers, which allow rather small successive changes in the controllers parameters.

To have a more accurate modeling and a better control performance, a new adaptive control scheme from nature responses such as a step-point response and a load disturbance response is presented here. Auto-tuning is used only to start up the controller. The on-line adaptive controller will wait for a significant output response to occur. If the response is due to a set-point change, then the identification method in Section 4.5 can be exploited to update the controller. This is straightforward and no further discussion will be given here. It should be pointed out that a set-point change is seldom made in process industry and control adaptation from a load disturbance response is much more demanding. If there is no set-point change, then any output response must be caused by some load disturbance. The problem is how to use the resultant transients to track the process frequency response and update the controller for the perturbed process. Load disturbance is often of low frequency and step signals are theoretically used as prototype disturbance (Astrom and Hagglund, 1995). Hang et al. (1995) have discussed the adaptation from step load disturbance response. However, in practice a load disturbance is seldom a pure step signal. In an adaptive method presented in this section, the load disturbance is not limited to a step signal, it can be a signal that is generated by inputting a step or an impulse through an unknown rational function dynamics and thus the new method can work with quite a large range of load disturbances.

4.7.1 Process estimation from load disturbance response

Consider a system shown in Fig. 4.29, where $G_c(s)$ is a controller and $G_p(s)$ is a process. d and n are load disturbance and measurement noise respectively.

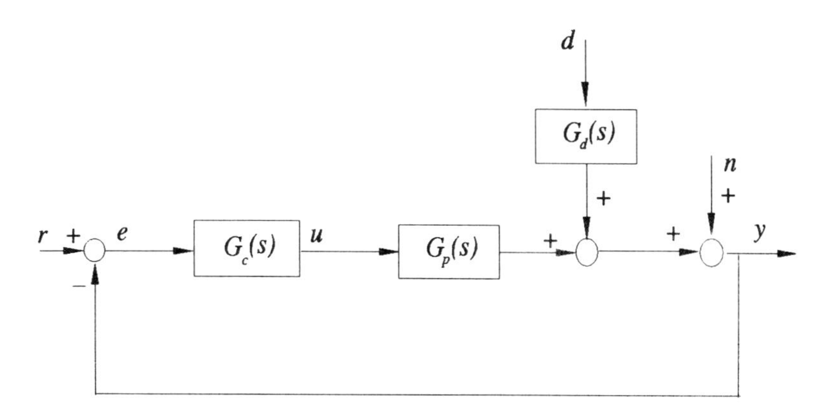

Fig. 4.29. Control system

The reason why the system configuration in Fig. 4.29 is adopted is that the new method to be described below employs only information on $u(t)$ and $y(t)$ to update process estimation and is thus general and applicable to different types of control schemes. Fig. 4.29 enables us to deal with one element G_c only and thus greatly simplifies the presentation. One may regard G_c as an equivalent single-loop controller if a control scheme other than single-loop one is actually in use.

Suppose that the system has been in its steady state. A process transient may be caused by set-point change or load disturbance. In the case of set-point change, normal process estimation methods can be employed to re-establish the process model. However, in industrial practice, set-point is often kept constant. If no set-point change has been made, any significant transient must be the result of some load disturbance. Load disturbance is often encountered in process control and the adaptation from load disturbance is thus attractive. However, an inherent property of disturbance is that they cannot be predicted exactly, they are usually modeled in some prototype disturbances: impulse, step, ramp, sinusoid (Astrom and Hagglund, 1995). The impulse and step like disturbances are commonly encountered in practice and will be the subject of study.

These disturbances can be regarded collectively as an equivalent disturbance d acting at the process output y through an unknown dynamic element G_d, as shown in Fig. 4.29. G_d is free of dead-time, since $G_d e^{-L_d s} d$ can always be transferred to $G_d \tilde{d}$, where $\tilde{d} = e^{-L_d s} d$ and G_d contains no dead-time. It should be pointed that d and G_d are unknown. Assume that a transient in the loop is caused by d and the resultant process responses $y(t)$ and $u(t)$ are recorded from the time when the $y(t)$ starts to change, to the time when the system settles down again. It is desired to re-identify the process G_p from the recorded responses so that the regulator G_c can be re- tuned and adapted to any change in G_p.

It follows from Fig. 4.29 that:

$$Y(s) = G_p(s)U(s) + G_d(s)D(s). \tag{4.134}$$

With $r = 0$, it follows that:

$$U(s) = -G_c(s)Y(s). \tag{4.135}$$

and

$$\frac{Y(s)}{D(s)} = \frac{G_d(s)}{1 + G_p(s)G_c(s)}. \tag{4.136}$$

Since d is unmeasurable, the new steady state of the process input may be checked and inferred from

$$D(s) = \begin{cases} 1, \ if \ u(\infty) = u(0); \\ \frac{1}{s}, \ if \ u(\infty) \neq u(0). \end{cases} \tag{4.137}$$

For $Y(s)$ in (4.136), $Y(s)$ can be calculated at $s = j\omega$ as follows. Suppose, without loss of generality, that the system is in the zero operating point before the load disturbance occurs and the controller contains integrator. $y(t)$ will return to its initial value after transient and $Y(j\omega)$ can be computed using Fast Fourier Transformation (FFT) of $y(t)$ (Hang et al., 1995). With $Y(j\omega)$ calculated and $D(j\omega)$ given in (4.137), it is then desired to be able to uniquely determine $G_p(j\omega)$ and $G_d(j\omega)$ from (4.136). The following example will illustrate the problem.

Example 4.21:

Suppose $G_c(s) = 1$, $G_p(s) = \frac{1}{s+2}$, $G_d(s) = \frac{1}{s+4}$, (4.136) becomes

$$\frac{Y(s)}{D(s)} = \frac{s+2}{(s+3)(s+4)}.$$

Suppose $G_c(s) = 1$, $G_p = \frac{2}{s+2}$, $G_d = \frac{1}{s+3}$, (4.136) also gives

$$\frac{Y(s)}{D(s)} = \frac{s+2}{(s+3)(s+4)}.$$

The example shows that even with the same $Y(s)$ and $D(s)$, $G_p(s)$ and $G_d(s)$ cannot be uniquely determined at the same time. However, under certain conditions, $G_p(s)$ and $G_d(s)$ can be solitarily determined.

Lemma 4.1:

If a process $G_p(s)$ has some dead-time while $G_d(s)$ has not, then with $Y(s)$ and $D(s)$ known, $G_p(s)$ and $G_d(s)$ can be uniquely determined.

Proof:

Suppose that besides $G_p(s) = G_r e^{-Ls}$ and $G_d(s)$, $\tilde{G}_p(s) = \tilde{G}_r(s)e^{-\tilde{L}s}$ and $\tilde{G}_d(s)$ also satisfy (4.136). From (4.136), it follows that

$$\frac{1}{G_d(s)} + \frac{G_c(s)}{G_d(s)}G_r(s)e^{-Ls} = \frac{1}{\tilde{G}_d(s)} + \frac{G_c(s)}{\tilde{G}_d(s)}\tilde{G}_r(s)e^{-\tilde{L}s}. \tag{4.138}$$

The first items on both hand-sides of (4.138) are dead-time free parts while the second item contain dead-time. To ensure that at all $s = j\omega$, $\omega \in R$, (4.138) is valid, the dead-time free parts on both hand-sides should be equal. So will the parts containing dead-time. Thus, one sees that

$$\begin{cases} G_d(s) = \tilde{G}_d(s), \\ G_r(s) = \tilde{G}_r(s), \\ L = \tilde{L}, \end{cases}$$

Hence the result.

It is fortunate that a large number of processes in process control contain some dead- time and thus $Y(s)$ and $D(s)$ can be used to calculate $G_p(s)$ and $G_d(s)$ simultaneously. It is known (Hang, 1991; Halevi, 1991) that most industrial processes can be approximately described by a low order plus dead-time model as:

$$G_p(s) = \frac{\beta s + 1}{\alpha_1 s^2 + \alpha_2 s + \alpha_3} e^{-Ls}. \tag{4.139}$$

G_d is modeled as:

$$G_d(s) = \frac{\gamma s + 1}{\lambda_1 s^2 + \lambda_2 s + \lambda_3}. \tag{4.140}$$

which can represent both the monotonic and oscillatory load disturbance responses. It follows from (4.134) that

$$\frac{\beta s + 1}{\alpha_1 s^2 + \alpha_2 s + \alpha_3} e^{-Ls} U(s) + \frac{\gamma s + 1}{\lambda_1 s^2 + \lambda_2 s + \lambda_3} D(s) = Y(s). \tag{4.141}$$

Estimating the parameters in (4.141) is not an easy job. However, (4.141) can be transfered (Whitfield, 1986) to

$$\begin{aligned} &a_1 s^4 Y(s) + a_2 s^3 Y(s) + \cdots + \lambda_3 Y(s) \\ &= b_1 s^3 U(s) e^{-Ls} + \cdots + b_3 s U(s) e^{-Ls} + \frac{\lambda_3}{\alpha_3} U(s) e^{-Ls} \\ &\quad + c_1 s^3 D(s) + \cdots + c_3 s D(s) + D(s). \end{aligned} \tag{4.142}$$

At $s = j\omega$, (4.142) is a linear equation in term of [$a_1, \ldots, a_4$, λ_3, b_1, b_2, b_3, $\frac{\lambda_3}{\alpha_3}$, c_1, c_2, c_3], where $Y(j\omega)$ and $D(j\omega)$ are known and $U(j\omega)$ in (4.142) is calculated using the method in Hang et al. (1995), which is shown below. The time response of process input $u(t)$ is decomposed into

$$u(t) = u(\infty) + \Delta u(t), \tag{4.143}$$

where $u(\infty)$ is the steady state, and $\Delta u(t)$ is the transient response. It follows that the process frequency response is

$$U(j\omega) = \frac{u(\infty)}{j\omega} + \Delta U(j\omega), \tag{4.144}$$

where $\Delta U(j\omega)$ can be estimated using Fast Fourier Transformation (FFT) of $\Delta u(t)$. Then, (4.142) is solved with Least Squares method. (4.142) can be further simplified, since $\frac{\lambda_3}{\alpha_3}$ in (4.142) can be directly calculated from the known $u(t)$, $y(t)$ and the assuming $d(t)$.

Lemma 4.2:

If the system is stable, $G_c(s)$ has an integrator, and d is a step disturbance, then $\frac{\lambda_3}{\alpha_3} = -\frac{1}{u(\infty)}$.

Proof:

Applying the Final Value Theorem to (4.141) gives

$$\frac{1}{\alpha_3} u(\infty) + \frac{1}{\lambda_3} d(\infty) = y(\infty).$$

Due to the integrator in the controller $G_c(s)$, $y(\infty) = 0$, and $d(\infty)$ is assumed to be one by (4.137). So $\frac{\lambda_3}{\alpha_3} = -\frac{1}{u(\infty)}$.

Lemma 4.3:

If the system is stable, $G_c(s)$ has an integrator, and d is an impulse disturbance, then $\frac{\lambda_3}{\alpha_3} = -\frac{1}{U(0)}$.

Proof:

Substituting $s = 0$ into (4.136) yields

$$\frac{Y(0)}{D(0)} = \frac{G_d(0)}{1 + G_p(0)G_c(0)}. \tag{4.145}$$

where $D(0) = 1$, $G_d(0) = \frac{1}{\lambda_3}$, $G_p(0) = \frac{1}{\alpha_3}$ and $G_c(0) = \infty$ due to the integral in controller $G_c(s)$. It is obvious from (4.145) that $Y(0) = 0$. With this result, substituting $s = 0$ into (4.141) gives

$$\frac{\lambda_3}{\alpha_3} = -\frac{1}{U(0)}.$$

Lemma 4.2 and Lemma 4.3 reduce the parametric estimation work. (4.142) can be rearranged into

$$\begin{aligned} & a_1 s^4 Y(s) + a_2 s^3 Y(s) + \cdots + a_5 Y(s) \\ & = b_1 s^3 U(s) e^{-Ls} + \cdots + b_3 s U(s) e^{-Ls} \\ +c_1 s^3 D(s) + \cdots + c_3 s D(s) + & [D(s) + \frac{\lambda_3}{\alpha_3} U(s) e^{-Ls}]. \end{aligned} \tag{4.146}$$

where only 11 coefficients need to be determined.

(4.146) is re-written as

$$\Phi X = \Gamma, \tag{4.147}$$

where $\Phi = [s^4Y(s),\ s^3Y(s), \cdots, Y(s),\ -s^3U(s)e^{-Ls}, \cdots,\ -sU(s)e^{-Ls},\ -s^3D(s),\ \cdots,\ -sD(s)]$, $\Gamma = D(s) + \frac{\lambda_3}{\alpha_3}U(s)e^{-Ls}$ and $X = [a_1, a_2, \ldots, a_5, b_1, b_2, b_3, c_1, c_2, c_3]^T$ are the real parameters to be estimated. Assume first that the process dead-time L is known, then with frequency responses $Y(j\omega_i)$, $U(j\omega_i)$ and $D(j\omega_i)$, $i = 1, 2, \cdots, m$, computed, (4.147) yields a system of linear algebraic equations at $s = j\omega_i$, $i = 1, 2 \cdots, m$. The least square solution X can be obtained in (4.147). The process frequency response $G_p(j\omega)$ can then be computed by

$$G_p(j\omega) = \frac{b_1 s^3 + b_2 s^2 + b_3 s + \frac{\lambda_3}{\alpha_3}}{a_1 s^4 Y + a_2 s^3 + a_3 s^2 + a_4 s + a_5} e^{-Ls}. \tag{4.148}$$

This solution in fact depends on L if L is unknown. The fitting error for (4.147) is given by

$$J(L) = \|\Phi(\Phi^T\Phi)^{-1}\Phi^T\Gamma - \Gamma\|_2, \tag{4.149}$$

which is a scalar nonlinear algebraic equation in one unknown L only. The error is then minimized with respect to L in the given interval, which is an iterative problem on one parameter L. Each iteration needs to solve a Least Squares problem corresponding to a particular value of L . The model parameters are determined when the minimum J is reached. To facilitate the solution further, some bounds for L will be derived so that the search can be constrained to a small interval. This will greatly reduce computations, improve numerical property, and produce a unique solution. It is noted that the phase lag contributed by the rational part of the model, $G_r(s) = \frac{\beta s+1}{\alpha_1 s^2+\alpha_2 s+\alpha_3}$, is bounded as

$$\arg G_r(j\omega) \in [-\pi, \frac{\pi}{2}], \quad \forall\omega \in (0, \infty), \tag{4.150}$$

so that an upper bound $\tilde{L}$ and a lower bound $\underline{L}$ can be imposed on L:

$$\tilde{L} = \min\left\{ -\frac{\arg G_p(j\omega_k) - \frac{\pi}{2}}{\omega_k} \right\}, \quad k = 1, 2, \cdots, m, \tag{4.151}$$

and

$$\underline{L} = \max\left\{ -\frac{\arg G_p(j\omega_k) + \pi}{\omega_k} \right\}, \quad k = 1, 2, \cdots, m. \tag{4.152}$$

Actually, from a relay feedback or set-point change, one can directly find out a gross estimate for dead-time by measuring the time $\tilde{L}$ between the control signal change to the output starting to move. Another possible bound may then be

$$L \in \left[0.5\tilde{L} \;\; 1.5\tilde{L}\right]. \tag{4.153}$$

Extensive simulation and real-time experiments show that within a reasonable bound, (4.149) exhibits a concave relationship with respect to L and yields a unique solution. Once L is determined, X is computed from (4.147).

Remark 4.7:

If the process dead-time L is unchanged since the last identification of G_p, no iteration is needed to solve (4.147). This is a special case that the bound for L is specified as a zero interval. This case greatly simplifies the identification. It may be true in many practical cases as process dynamics perturbations are usually associated with operating point changes and/or load disturbance, which mainly cause time constant/gain changes. Furthermore, any small dead-time change can be discounted in other parameter changes.

Remark 4.8:

Weighting can be considered when solving (4.147). Distortions caused by transferring a nonlinear problem of (4.141) to a linearized problem of (4.147) can be reduced using weighting (Whitfield, 1986). To have a better result, more weights should be given to the frequency region where $Y(j\omega)$ and $U(j\omega_i)$ has less computation error. This is especially important when real-time data is considered.

Remark 4.9:

The measurement noise n in Fig. 4.29 is usually of high frequency while the process frequency response of interest for control analysis and design is usually in the low frequency region. In particular, the process frequency response from 0 to the critical frequency ω_π is mostly critical for Controller Design. It is observed that in the experiments the measurement noise is indeed in the fairly high frequency region. Therefore, a low pass filter can be employed to reduce the measurement noise. The cut-off frequency of the filter is determined with respect to the process frequency region of interest. A possible choice is $(3 \sim 5)\omega_\pi$.

Remark 4.10:

If the set-point is not changed after the system enters its steady state, the transients in process input and output may also be caused by the process static gain change. The process response under this change is equivalent to the process response under a load disturbance. If the process gain abruptly changes to its final value, then the process estimation from load disturbance response is still valid. This can be shown by Fig. 4.30.

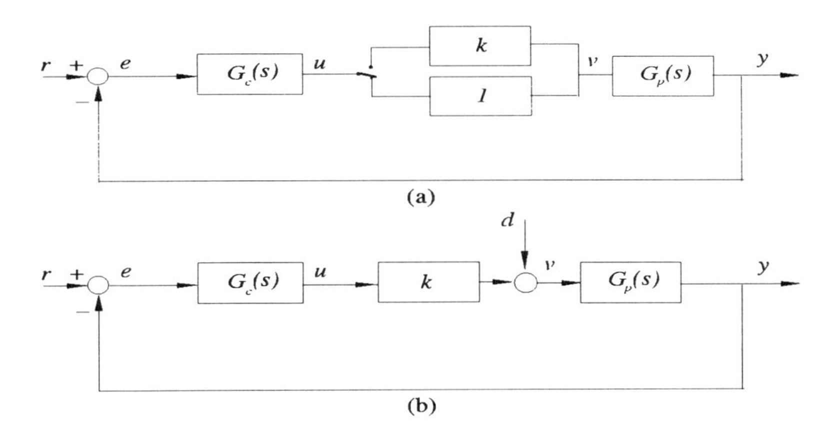

Fig. 4.30. Equivalent systems when gain changes.

Suppose that at $t = t_0^-$, the systems in Fig. 4.30(a) and Fig. 4.30(b) have both entered the steady states. At $t = t_0$, the process gain in Fig. 4.30(a) suddenly jumps from 1 to k. It follows that $\nu(t_0^-) = u_0$, $\nu(t_0^+) = ku_0$ and $y(t_0^-) = y_0$, where u_0 and y_0 are the steady states. The system in Fig. 4.30(b) before $t = t_0$ is also in steady state with $d = -(k-1)u_0$. At $t = t_0$ and afterwards, the load disturbance changes to zero. From Fig. 4.30(b), it also follows that $\nu(t_0^-) = u_0$, $\nu(t_0^+) = ku_0$ and $y(t_0^-) = y_0$. From $t = t_0$ onwards, the systems in Fig 4.30(a) and Fig. 4.30(b) have the same structure and parameters, and the initial values at each point in the two plots are the same. Therefore, the systems in Fig. 4.30(a) and Fig. 4.30(b) have the same responses after $t = t_0$ and they are equivalent. Fig. 4.30(b) can be further modified to Fig. 4.29, and the estimation method can be applied. This is demonstrated by an example later on.

4.7.2 PID adaptation

After the process frequency response $G_p(j\omega_i)$, $i = 1, 2, \cdots, m$, has been re-identified, a PID controller:

$$G_c(s) = K_p\left(1 + \frac{1}{T_i s} + T_d s\right),$$

may be re-tuned by matching $G_p(j\omega_i)G_c(j\omega_i)$ to the desired open-loop $G_{yr}(j\omega_i)$. The desired closed-loop transfer function $\bar{G}_{yr}$ is set (Wang et al., 1997) as

$$\bar{G}_{yr}(s) = \frac{\omega_n^2}{s^2 + 2\zeta\omega_n s + \omega_n^2} e^{-Ls}$$

according to the given specifications. If the control specifications are not given by the user, then the default settings can be used for the parameter ζ and $\omega_n L$, and preferably they are $\zeta = 0.707$ and $\omega_n L = 2$, which implies that the overshoot of the objective set-point step response is about 5%, the phase margin is 60° and the gain margin is 2.2. The desired open-loop $G_{yr}(j\omega)$ is thus given by:

$$G_{yr}(j\omega_i) = \frac{\bar{G}_{yr}(j\omega_i)}{1 - \bar{G}_{yr}(j\omega_i)}.$$

The PID controller parameters can be obtained with linear least squares method (Wang et al., 1997). The controller is then able to adapt to the possible changes in process.

Several typical processes are employed in demonstrating the method. For assessment of accuracy, the identification error is here measured by worst case error

$$ERR = \max_i \left\{ \left| \frac{\tilde{G}_p(j\omega_i) - G_p(j\omega_i)}{G_p(j\omega_i)} \right| \times 100\%, i = 1, 2, \ldots, M \right\}, \quad (4.154)$$

where $G_p(j\omega_i)$ and $\tilde{G}_p(j\omega_i)$ are the actual and the estimated process frequency responses respectively. The Nyquist curve for phase ranging from 0 to $-\pi$ is being considered since this part is most significant for control design.

Example 4.22:

Consider the control system in Fig. 4.29. Suppose initially that the system is in the steady state and for some reason the process is changed to

$$G_p(s) = \frac{1}{s^2 + 3s + 2} e^{-s}.$$

Thereafter, a step disturbance comes into the system through an unknown dynamics:

$$G_d(s) = \frac{1}{s+3}.$$

And the significant transients in the system are detected. The process output $y(t)$ and input $u(t)$ are then logged until the process settles down again. $Y(j\omega)$ is computed using FFT, $U(j\omega)$ is determined using (4.144) while $D(j\omega)$ is taken as step-type since the steady state of $u(t)$ reaches a new value. The process frequency response is calculated by (4.147) and (4.148), which is shown in Fig. 4.31.

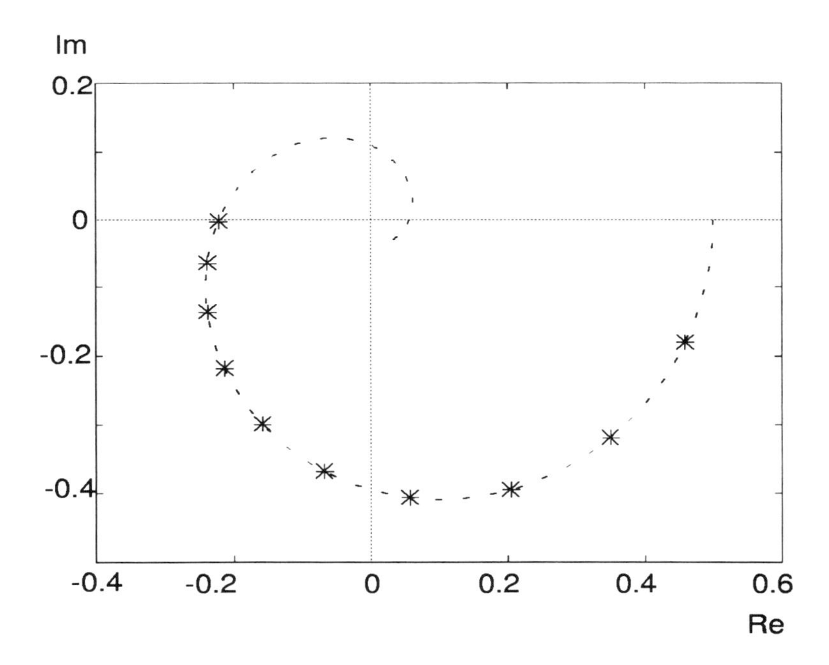

Fig. 4.31. Nyquist plot. $\cdots + \cdots$ Actual, $\times$ Estimated

The ERR is 0.47%. This indicates that the method provides a very accurate process frequency response.

A step disturbance may come through different unknown dynamics. Suppose that they are respectively $G_d(s) = \frac{1}{s^2+4s+3}$ and $G_d(s) = \frac{1}{s^2+5s+10}$ and the identification errors are listed in Table 4.5. An impulse input is also considered, the re-identification results are also very satisfactory, as shown in Table 4.5.

Example 4.23:

Suppose that a second order oscillatory process has perturbed to

$$G_p(s) = \frac{1}{s^2 + s + 1} e^{-2s}.$$

From the step/pulse-like load disturbance responses, the process can be re-estimated. Different load disturbances generated by passing a step or an impulse through different unknown dynamics are employed to illustrate the adaptation method. The estimation errors are listed in Table 4.5.

Table 4.5. Identification error from load disturbance

$G_p(s)$	$\frac{1}{s^2+3s+2}e^{-s}$		$\frac{1}{s^2+s+1}e^{-2s}$		$\frac{1}{(s+1)^5}e^{-4s}$	
$G_d(s)$	Step	Pulse	Step	Pulse	Step	Pulse
$\frac{1}{s+3}$	0.47%	0.13%	3.85%	0.72%	1.33%	0.44%
$\frac{1}{s^2+4s+3}$	3.30%	1.21%	5.51%	0.59%	1.24%	0.95%
$\frac{3}{s^2+5s+10}$	1.40%	0.75%	0.29%	0.51%	3.86%	1.37%

Example 4.24:

For a high order process:

$$G_p(s) = \frac{1}{(s+1)^5} e^{-4s},$$

it is re-identified under different load disturbance cases. The estimation errors ERR are given in Table 4.5.

The above examples are of noise free cases. The estimation results under noise pollution are also studied, where the noise source is a white noise. The noise-to-signal ratio (Haykin, 1989) is defined as Noise-to-Signal Power Spectrum Ratio N_1. The low pass filter is selected as a Butterworth low pass filter whose cut-off frequency is chosen as $3 \sim 5\omega_\pi$. The process in Example 4.24 is used to illustrate the results under different noise levels, where the unknown disturbance is obtained by passing a step signal through $\frac{1}{s^2+4s+3}$. The estimation errors are shown in Table 4.6, which are quite satisfactory.

Table 4.6. Estimation error under different noise levels

N_1	0	4.50%	9.00%	17.5%	25.5%
ERR	1.24%	3.11%	5.91%	8.64%	12.33%

To test the robustness of the method, a non-step/impulse input is applied to the unknown element $G_d(s)$. For a process:

$$G_p(s) = \frac{1}{s^2 + 3s + 2} e^{-s}.$$

a non-step disturbance shown in Fig. 4.32(a) is introduced to the system through $G_d(s) = \frac{1}{s+3}$. The non-step input is still treated as a step-input and the same re-identification procedure conducted. The resultant error ERR is 9.82%. A non-impulse input as in Fig. 4.32(b) is also employed to show the robustness of the method. The process frequency response estimation error ERR is 9.33%.

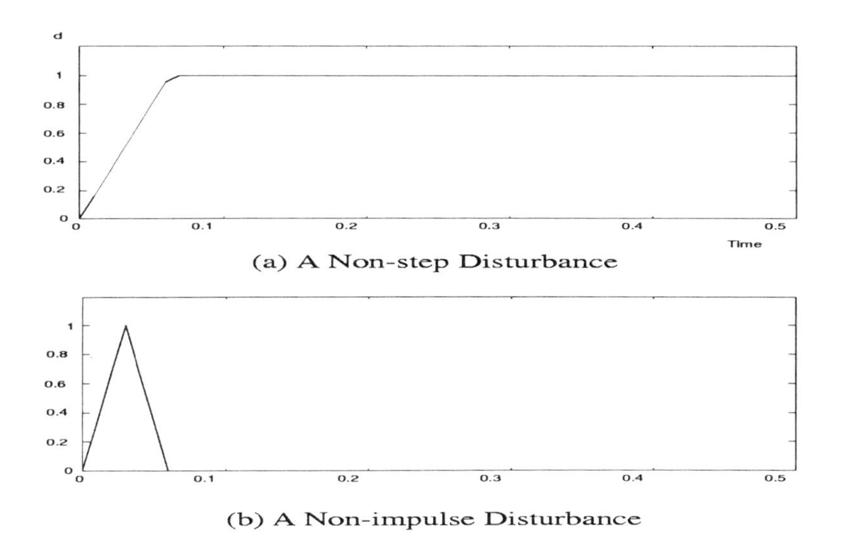

Fig. 4.32. Disturbance signal for robustness testing

Real-time Experiment. The method was tested on the Dual Process Simulator KI 100. To observe the noise effect in the test, apart from the inherent test environment noise, extra noise with peak-to-peak value of 0.1 V is introduced using the noise source in the Simulator.

As a demonstration, the process is chosen from the simulator as

$$G_p(s) = \frac{1}{(5s+1)^2} e^{-5s}.$$

With the PID design method (Wang et al., 1997), the PID controller is obtained as

$$G_c(s) = 0.66 \left(1 + \frac{1}{10.04s}\right). \tag{4.155}$$

The process response to a set-point change at $t = 0$ is pretty good, which is shown in Fig. 4.33. Suppose that after the process settles down, the process gain suddenly changes from one to two at $t = 66s$ as

$$G_p(s) = \frac{2}{(5s+1)^2} e^{-5s}.$$

The transients of the process input and output are recorded as shown in Fig. 4.33. As shown before, the process response under gain change is equivalent to a step load disturbance. Thus, the process model can be estimated using the method presented in Section 4.7.1 and the PID controller is re-tuned accordingly. This results in a new controller:

$$G_c(s) = 0.45 \left(1 + \frac{1}{11.67s} + 0.11s\right). \tag{4.156}$$

The next step set-point change at $t = 228s$ leads to a quite satisfactory response, as shown in Fig. 4.33.

For comparison, if the controller in (4.155) was still used for the new process, the resultant response (dashed line) is also displayed in Fig. 4.33. Assume next that the process is further perturbed from second order to fourth order and dead-time changes to 2.5 as

$$G_p(s) = \frac{2}{(5s+1)^4} e^{-2.5s}.$$

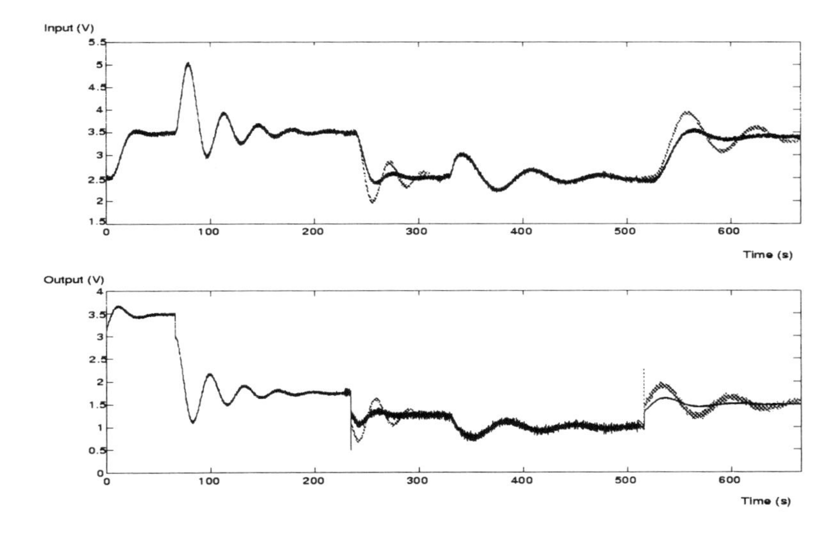

Fig. 4.33. Real-time adaptation test.

An unknown load disturbance occurs, which is generated by applying a step signal through an unknown disturbance channel:

$$G_d = \frac{1}{(2s+1)^2}.$$

The load disturbance response is used to run the adaptation scheme. The process is updated and the controller is adjusted to

$$G_c(s) = 0.33\left(1 + \frac{1}{15.23s}\right). \tag{4.157}$$

to adapt to the process perturbation. A set-point change occurs at $t = 510s$. The control performance of the new process under controller in (4.156) is shown in Fig. 4.33 and the adaptation performance by the method is also illustrated there. The real-time test shows that the method is capable of adapting the controller using a benign unknown load disturbance response.

CHAPTER 5
MULTI-LOOP CONTROL

5.1 Introduction

Processes with inherently more than one variable to be controlled are frequently encountered in the industries and they are known as multivariable or multi-input multi-output (MIMO) processes. Interactions usually exist between control loops, which account for the renowned difficulty in their control compared to single-input single-output (SISO) processes. The goal to achieve satisfactory loop performance has hence posed a great challenge in the area of control design. Depending on the application and requirement, either a fully cross-coupled or a multi-loop controller can be adopted for MIMO processes. Although multivariable controllers are capable of providing explicit suppression of interactions, their designs are usually more complicated and practical implementation inevitably more costly. Multi-loop controllers, sometimes known as decentralized controllers, have a simpler structure and, accordingly, less tuning parameters than the fully cross-coupled one. In addition, in the event of component failure, it is relatively easy to stabilize manually, since only one loop is directly affected by the failure (Palmor 1996; Skogestad and Morari 1989). Hence for processes with modest interactions, multi-loop controllers are often more favorable than multivariable controllers.

Many multi-loop controller design methods have been reported in the literature. In the BLT (Biggest Log Modulus Tuning) method (Luyben 1986), the familiar Ziegler-Nichols rule is modified with the inclusion of a detuning factor, which determines the tradeoff between stability and performance of the system. Individual controllers are designed for the respective loops by first ignoring all interactions. The calculated controller gains are then scaled by the detuning factor to guarantee stability. Despite the simple computations involved, the design views interactions as elements impeding system stability and attempts to suppress their effects rather than control them to speed up individual loops. Hence, it is rather conservative. In the sequential loop closing method, the loops are closed one after another, with those previously tuned closed with appropriate controllers. The main drawback of this method is that the design proceeds in a very *ad hoc* manner. Design decisions made when closing the first one or two loops may have undesirable effects on the behavior of the remaining loops. The interactions are well taken care of only if the loops are of considerably different bandwidths and the closing sequence begins from the fastest loop. These assumptions can rarely be justified (Maciehowski, 1986). Generally, the closing of a new loop will bring about interaction to all the previously closed loops.

In this chapter, a new method is presented for the design of multi-loop controllers, which does not suffer from the above drawbacks and avoids the conservativeness of the BLT method by handling all possible interactions from all loops simultaneously. It is a multi-loop extension of the Ziegler-Nichols frequency response method for SISO systems. In the face of loop interactions, a direct application of the method to the diagonal loop elements according to the SISO way may not be appropriate. To take into account the multivariable interactions, each loop is viewed as an independent equivalent process with all possible interactions lumped as one integrated process, and controllers are designed to meet the specifications that a given point on Nyquist curve be moved to a desired position for each equivalent process. A novel approach is presented to obtain the controller parameters which achieve the above design criterion. Simulation examples are given to show the effectiveness of the method and comparisons are made with the BLT method.

5.2 The Modified Ziegler-Nichols Method

In Chapter 2, it has been illustrated that the Ziegler-Nichols tuning methods may be classified under the more general frequency response shaping framework which is sometimes also referred to as the modified Ziegler-Nichols method (MZN). The method is briefly described as follows. Given a point A on the Nyquist curve of the process $g(s)$:

$$A \triangleq G_p(j\omega) = r_a e^{j(-\pi+\varphi_a)},$$

a controller $G_c(s)$ is to be computed to compensate for $G_p(s)$ such that this point is moved to:

$$B \triangleq G_p(j\omega)G_c(j\omega) = r_b e^{j(-\pi+\varphi_b)}. \tag{5.1}$$

When the controller is chosen as:

$$G_c(s) = K_c(1 + \frac{1}{T_i s} + T_d s),$$

a solution to (5.1) is

$$K_c = \frac{r_b \cos(\varphi_b - \varphi_a)}{r_a}, \tag{5.2}$$

$$\omega T_d - \frac{1}{\omega T_i} = \tan(\varphi_b - \varphi_a). \tag{5.3}$$

One notes that the controller gain K_c is uniquely given by (5.2), but T_i and T_d are not. An additional condition must thus be introduced to determine these two latter parameters uniquely. A common method is to specify a relation between T_i and T_d as:

$$T_d = \alpha T_i, \tag{5.4}$$

where α is recommended (Hagglund and Astrom 1991) to be equal to 0.25. Another method to specify α is given below by (5.10), which can generally yield better performance. In many practical cases such as high noise, it is desirable not to use derivative action and a PI controller:

$$G_c(s) = K_c(1 + \frac{1}{T_i s}).$$

is preferred. The PI controller can be viewed as a special case of a PID controller with $T_d = 0$, or $\alpha = 0$. It is clear that if a PI controller is adopted, $\varphi_b - \varphi_a$ should be chosen negative to make the specification achievable as a PI controller always gives phase lag. With (5.4), T_i is obtained from (5.3) as:

$$T_i = \frac{\beta}{\omega}, \tag{5.5}$$

where

$$\beta = \begin{cases} \frac{\tan(\varphi_b - \varphi_a) + \sqrt{4\alpha + \tan^2(\varphi_b - \varphi_a)}}{2\alpha}, & \alpha > 0, \\ -\frac{1}{\tan(\varphi_b - \varphi_a)}, & \alpha = 0, \end{cases}$$

and T_d is then given by (5.4).

It should be noted that the performance achieved from these tuning methods may vary significantly. This is not surprising as a critical point is moved to different positions in these methods. Besides, some of these tuning methods may not produce satisfactory closed-loop responses in certain circumstances. For instance, the Ziegler-Nichols tuning laws usually result in rather oscillatory set point responses. It is thus important to know where the critical point should be moved to for acceptable performance with some prior knowledge of the process.

Zhuang and Atherton (Zhuang and Atherton 1993) considered the tuning of PI/PID controller to achieve optimum integral time-weighted square error (ITSE) performance criterion. The optimum tuning method can be interpreted as the movement of the critical point to a desired position specified by:

$$\varphi_b = -\arctan(\frac{1}{0.166\pi(1.935\kappa + 1)}), \quad (5.6)$$

$$r_b = 0.361/\cos\varphi_b, \quad (5.7)$$

for the PI controller, or

$$\varphi_b = 0.59(1 - 0.97e^{-0.45\kappa}), \quad (5.8)$$

$$r_b = 0.614(1 - 0.233e^{-0.347\kappa}), \quad (5.9)$$

for the PID controller, where κ is the process normalized gain:

$$\kappa = \left| \frac{G_p(j0)}{G_p(j\omega_c)} \right|,$$

and ω_c is the process critical frequency. The parameter α for PI and PID controllers is given by $\alpha = 0$ and

$$\alpha = \frac{0.413}{3.302\kappa + 1}, \quad (5.10)$$

respectively.

According to extensive simulation and experiments, the modified Ziegler-Nichols method plus the above optimum choice of the desired point B provides the best performance among all the existing methods using only information relating to the steady state gain and the critical points of the process. In what follows, the MZN will be extended to the multi-loop case for the control of a multivariable process, where the use of (5.6) to (5.10) is optional but recommended.

5.3 Review of the BLT (Biggest Log-Modulus Tuning)

Before proceeding to develop the multi-loop version of the modified Ziegler-Nichols method, the BLT (Biggest Log-Modulus Tuning) (Luyben 1986) will be first reviewed. The BLT method is a widely used method as it provides a standard tuning methodology for a multi-loop controller. It allows reasonable controller settings to be obtained with only modest engineering and computational effort. It is not claimed that the method will produce the best performance, but one which provides reasonable settings for further fine tuning, and also serves as a benchmark for comparative studies. BLT tuning involves the following four steps:

1. Calculate the ultimate gain and ultimate frequency of each diagonal transfer function $g_{i_1 i_1}(s)$ and obtain the Ziegler-Nichols settings for each individual loop.

2. A detuning factor F is assumed. F should always be greater than 1. Typical values are $1.5 \sim 4$. The gains of all feedback controllers K_{c,i_1} are calculated by dividing the Ziegler-Nichols gains K_{ZN,i_1} by the factor F. Then all feeback controller reset time T_{i,i_1} are calculated by multiplying the Ziegler-Nichols reset times T_{ZN,i_1} by the same factor F. The factor can be considered as a detuning factor which is applied to all loops. The larger the value of F, the more stable the system will be at the expense of a more sluggish setpoint and load responses.

3. Using an initial value of F and the resulting controller settings, the multivariable Nyquist plot of the scalar function $W(j\omega) = -1 + |I + G_p(j\omega)G_c(j\omega)|$ is produced. The closer this contour is to the $(-1,0)$ point, the closer the system is to instability. Therefore, based on intuition and empirical grounds, a multivariable closed-loop log modulus $L_{cm} \triangleq 20log_{10}|\frac{W}{1+W}|$ can be defined. The peak in the plot of L_{cm} over the entire frequency range is defined as the biggest log modulus L_{cm}^{max}.

4. The F factor is varied until L_{cm}^{max} is equal to $2N$, where N is the order of the system. This empirically determined criterion has been tested on a large number of cases and it gives reasonable performance, although slightly on the conservative side.

5.4 Modified Ziegler-Nichols Method for Multi-Loop Processes

Consider a stable 2x2 process described via the transfer function matrix $G_p(s)$ as $Y(s) = G_p(s)U(s)$ or

$$\begin{bmatrix} y_1(s) \\ y_2(s) \end{bmatrix} = \begin{bmatrix} g_{11}(s) & g_{12}(s) \\ g_{21}(s) & g_{22}(s) \end{bmatrix} \begin{bmatrix} u_1(s) \\ u_2(s) \end{bmatrix}.$$

Assume that proper input-output pairing has been made to the process. If the process is inherently poorly paired, the RGA (Relative Gain Array) method (Maciehowski 1989) may be employed to make possible the necessary arrangement. The process is to be controlled in a negative feedback configuration by the multi-loop controller:

$$G_c(s) = \begin{bmatrix} k_1(s) & 0 \\ 0 & k_2(s) \end{bmatrix}.$$

The resultant control system is shown in Figure 5.1. The controller design objective is to find $G_c(s)$ such that both loops achieve satisfactory performance.

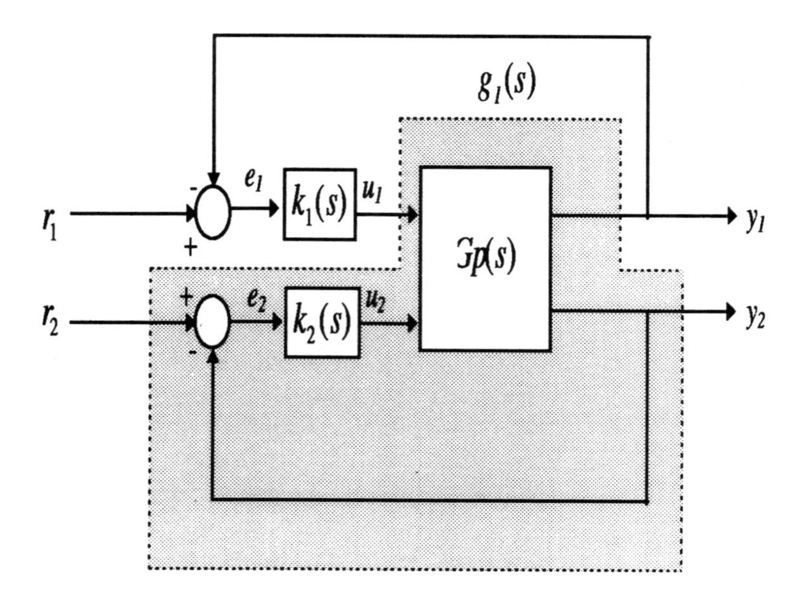

Fig. 5.1. Equivalent $g_1(s)$ for a 2×2 process

The boxed portion in Figure 5.1 can be interpreted as an individual SISO process with an equivalent transfer function $g_1(s)$ between input u_1 and output y_1. It follows that $g_1(s)$ can be obtained (Maciehowski 1989) as:

$$g_1 = g_{11} - \frac{g_{12}g_{21}}{k_2^{-1} + g_{22}}. \tag{5.11}$$

Similarly, the equivalent process between u_2 and y_2 is given by:

$$g_2 = g_{22} - \frac{g_{21}g_{12}}{k_1^{-1} + g_{11}}. \tag{5.12}$$

It is thus clear that in order to achieve good loop performance, $k_1(s)$ and $k_2(s)$ should be designed for $g_1(s)$ and $g_2(s)$ instead of the process diagonal element $g_{11}(s)$ and $g_{22}(s)$ which is the case in many other methods (Luyben 1986). In the sequential loop closing method (Maciehowski 1989), $k_1(s)$ is first designed for $g_{11}(s)$ (assuming that loop closing starts from loop one). Subsequently, $k_2(s)$ is designed for the equivalent transfer function $g_2(s)$. While the performance of the second loop will be consistent with what is expected from the design, the performance of the first one will not, as the closing of the second loop will bring about its interactions into the first loop which has not been taken into account when $k_1(s)$ is designed.

In what follows, the modified Ziegler-Nichols method described in the preceding section is applied to the equivalent transfer function $g_1(s)$ and $g_2(s)$, i.e., the controller $k_{i_1}(s)$, $i_1 = 1,2$, are designed such that a given point on the Nyquist curve of $g_{i_1}(s)$, $i_1 = 1,2$, is moved to a desired point. The selection of these two points is recommended to follow the same rules in the preceding section. In this way, the optimum settings for the multi-loop controller can be obtained with all the benefits derived from the well developed SISO tuning. Let

$$A_{i_1} = g_{i_1}(j\omega_{i_1}) = r_{a,i_1} e^{j(-\pi+\varphi_{a,i_1})}, \qquad i_1 = 1,2, \tag{5.13}$$

be the two given points on the Nyquist curves of $g_{i_1}(s)$, $i_1 = 1,2$. They are to be moved respectively to the desired points

$$B_{i_1} = g_{i_1}(j\omega_{i_1}) k_{i_1}(j\omega_{i_1}) = r_{b,i_1} e^{j(-\pi+\varphi_{b,i_1})}, \qquad i_1 = 1,2. \tag{5.14}$$

It should be pointed out that, unlike the SISO case, due to the dependence of g_1 (or g_2) on k_2 (or k_1), ω_1 and thus k_1 (or ω_2 and k_2) cannot be determined until k_2 (or k_1) has been fixed. This results in a cause-effect loop and causes a major design difficulty. Let $k_1(s)$ and $k_2(s)$ be of PID type, i.e.

$$k_{i_1}(s) = K_{c,i_1}(1 + \frac{1}{T_{i,i_1} s} + T_{d,i_1} s), \qquad i_1 = 1,2,$$

which reduces to PI type when $T_{d,i_1} = 0$. Now the modified Ziegler-Nichols method specified in (5.2), (5.4) and (5.5) is applied to each $k_{i_1}(s)$. Invoking (5.4) and (5.5) to each $k_{i_1}(s)$ yields:

$$k_{i_1}(s) = K_{c,i_1}\left(1 + \frac{\omega_{i_1}}{\beta_{i_1} s} + \alpha_{i_1}\frac{\beta_{i_1}}{\omega_{i_1}} s\right), \qquad i_1 = 1,2, \tag{5.15}$$

where

$$\beta_{i_1} = \begin{cases} \frac{\tan(\varphi_{b,i_1}-\varphi_{a,i_1}) + \sqrt{4\alpha_{i_1} + \tan^2(\varphi_{b,i_1}-\varphi_{a,i_1})}}{2\alpha_{i_1}}, & \alpha_{i_1} > 0, \\ -\frac{1}{\tan(\varphi_{b,i_1}-\varphi_{a,i_1})}, & \alpha_{i_1} = 0. \end{cases} \tag{5.16}$$

$\alpha_{i_1} > 0$ results in full PID control, and $\alpha_{i_1} = 0$ reduces to PI control. However, (5.2) cannot be used directly because $K_{c,1}$ is related to $r_{a,1}$ which depends on the unknown $K_{c,2}$. It follows from the phase part of (5.14) that

$$k_{i_1}(j\omega_{i_1}) = K_{c,i_1}(1 + j\tan(\varphi_{b,i_1} - \varphi_{a,i_1})), \qquad i_1 = 1,2.$$

(5.14) can be written as:

$$g_{i_1}(j\omega_{i_1}) K_{c,i_1} = \cos(\varphi_{b,i_1} - \varphi_{a,i_1}) r_{b,i_1} e^{j(-\pi+\varphi_{a,i_1})}.$$

It follows from (5.11), (5.12) and (5.15) that

$$\{g_{11}(j\omega_1) - \frac{g_{12}(j\omega_1)g_{21}(j\omega_1)}{\frac{1}{K_{c,2}[1+j(\alpha_2\frac{\beta_2\omega_1}{\omega_2}-\frac{\omega_2}{\beta_2\omega_1})]} + g_{22}(j\omega_1)}\}K_{c,1}$$
$$= \cos(\varphi_{b,1} - \varphi_{a,1})r_{b,1}e^{j(-\pi+\varphi_{a,1})},$$

$$\{g_{22}(j\omega_2) - \frac{g_{12}(j\omega_2)g_{21}(j\omega_2)}{\frac{1}{K_{c,1}[1+j(\alpha_1\frac{\beta_1\omega_2}{\omega_1}-\frac{\omega_1}{\beta_1\omega_2})]} + g_{11}(j\omega_2)}\}K_{c,2}$$
$$= \cos(\varphi_{b,2} - \varphi_{a,2})r_{b,2}e^{j(-\pi+\varphi_{a,2})},$$

or

$$\frac{g_{11}(j\omega_1)}{[1+j(\alpha_2\frac{\beta_2\omega_1}{\omega_2}-\frac{\omega_2}{\beta_2\omega_1})]}K_{c,1} + \cos(\varphi_{b,1} - \varphi_{a,1})r_{b,1}e^{j(-\pi+\varphi_{a,1})}$$
$$g_{22}(j\omega_1)K_{c,2} + \Delta_1 K_{c,1}K_{c,2} = \frac{\cos(\varphi_{b,1} - \varphi_{a,1})r_{b,1}e^{j(-\pi+\varphi_{a,1})}}{[1+j(\alpha_2\frac{\beta_2\omega_1}{\omega_2}-\frac{\omega_2}{\beta_2\omega_1})]}, \quad (5.17)$$

$$\frac{g_{22}(j\omega_2)}{[1+j(\alpha_1\frac{\beta_1\omega_2}{\omega_1}-\frac{\omega_1}{\beta_1\omega_2})]}K_{c,2} + \cos(\varphi_{b,2} - \varphi_{a,2})r_{b,2}e^{j(-\pi+\varphi_{a,2})}$$
$$g_{11}(j\omega_2)K_{c,1} + \Delta_2 K_{c,2}K_{c,1} = \frac{\cos(\varphi_{b,2} - \varphi_{a,2})r_{b,2}e^{j(-\pi+\varphi_{a,2})}}{[1+j(\alpha_1\frac{\beta_1\omega_2}{\omega_1}-\frac{\omega_1}{\beta_1\omega_2})]}, \quad (5.18)$$

where

$$\Delta_1 = g_{11}(j\omega_1)g_{22}(j\omega_1) - g_{12}(j\omega_1)g_{21}(j\omega_1),$$
$$\Delta_2 = g_{11}(j\omega_2)g_{22}(j\omega_2) - g_{12}(j\omega_2)g_{21}(j\omega_2).$$

Since φ_{a,i_1}, φ_{b,i_1}, r_{b,i_1}, α_{i_1} are to be specified, there are altogether four unknowns, namely, $K_{c,1}$, $K_{c,2}$, ω_1 and ω_2 in equation (5.17) and (5.18). These two equations are complex and they can be broken down to four real equations. Since the number of unknowns equals the number of real equations, intuitively, it may appear that the problem can be easily solved. Unfortunately this is not the case. The main difficulty lies in the non-linearities of the equations which are further complicated by the coupling among them. In what follows, a novel procedure for solving the above equations is presented based on a graphical approach.

5.5 Derivation of the Design Equations

Since (5.17) is complex, it may be broken down into two real equations:

$$\mathrm{Re}[\Delta_1]K_{c,1}K_{c,2} + \mathrm{Re}[a_1]K_{c,1} + \mathrm{Re}[b_1]K_{c,2} + \mathrm{Re}[c_1] = 0,$$
$$\mathrm{Im}[\Delta_1]K_{c,1}K_{c,2} + \mathrm{Im}[a_1]K_{c,1} + \mathrm{Im}[b_1]K_{c,2} + \mathrm{Im}[c_1] = 0,$$

where

$$\begin{aligned} a_1 &= \frac{g_{11}(j\omega_1)}{[1 + j(\alpha_2 \frac{\beta_2 \omega_1}{\omega_2} - \frac{\omega_2}{\beta_2 \omega_1})]}, \\ b_1 &= -\cos(\varphi_{b,1} - \varphi_{a,1}) r_{b,1} e^{j(-\pi + \varphi_{a,1})} g_{22}(j\omega_1), \\ c_1 &= -\frac{\cos(\varphi_{b,1} - \varphi_{a,1}) r_{b,1} e^{j(-\pi + \varphi_{a,1})}}{[1 + j(\alpha_2 \frac{\beta_2 \omega_1}{\omega_2} - \frac{\omega_2}{\beta_2 \omega_1})]}. \end{aligned}$$

$K_{c,1}$ can be expressed in terms of $K_{c,2}$ using the first equation as:

$$K_{c,1} = -\frac{\mathrm{Re}[b_1]K_{c,2} + \mathrm{Re}[c_1]}{\mathrm{Re}[\Delta_1]K_{c,2} + \mathrm{Re}[a_1]}. \tag{5.19}$$

The expression is next substituted into the second equation to obtain

$$\begin{aligned} &(\mathrm{Re}[\Delta_1]\,\mathrm{Im}[b_1] - \mathrm{Im}[\Delta_1]\,\mathrm{Re}[b_1])\,K_{c,2}^2 \\ &\quad + (\mathrm{Re}[\Delta_1]\,\mathrm{Im}[c_1] - \mathrm{Im}[\Delta_1]\,\mathrm{Re}[c_1] + \mathrm{Re}[a_1]\,\mathrm{Im}[b_1] - \mathrm{Im}[a_1]\,\mathrm{Re}[b_1]) \\ &\qquad K_{c,2} + (\mathrm{Re}[a_1]\,\mathrm{Im}[c_1] - \mathrm{Im}[a_1]\,\mathrm{Re}[c_1]) = 0. \end{aligned} \tag{5.20}$$

The above equation is quadratic in $K_{c,2}$ and it has two solutions which must be verified for validity. First of all, only real solutions should be accepted for physical realizability. This requires

$$\begin{aligned} &(\mathrm{Re}[\Delta_1]\,\mathrm{Im}[c_1] - \mathrm{Im}[\Delta_1]\,\mathrm{Re}[c_1] + \mathrm{Re}[a_1]\,\mathrm{Im}[b_1] - \mathrm{Im}[a_1]\,\mathrm{Re}[b_1])^2 \\ &\quad - 4\,(\mathrm{Re}[\Delta_1]\,\mathrm{Im}[b_1] - \mathrm{Im}[\Delta_1]\,\mathrm{Re}[b_1]) \\ &\qquad (\mathrm{Re}[a_1]\,\mathrm{Im}[c_1] - \mathrm{Im}[a_1]\,\mathrm{Re}[c_1]) \geq 0. \end{aligned} \tag{5.21}$$

Since a_1, b_1, c_1 and Δ_1 are functions of ω_1 and ω_2, (5.21) restricts (ω_1, ω_2) to an admissible region, but the search for a solution set is still not a simple task. A suitable search strategy is to provide an initial guess $\omega_2^{(0)}$ on ω_2, and then find all ω_1 satisfying the condition (5.21). Such a ω_1 will be in a union of intervals. It will now be shown that not all the intervals are needed, and only one is sufficient for the design. Notice that when $K_{c,2} = 0$, i.e., the second loop is opened, $g_1(s) = g_{11}(s)$ according to (5.11). Following (5.13), ω_1 should be the ω_{11} satisfying:

$$\arg g_{11}(j\omega_{11}) = -\pi + \varphi_{a,1}. \tag{5.22}$$

By varying the controller gain $K_{c,2}$, a different amount of interactions is brought into the first loop and thus $g_1(s)$ also changes. When $K_{c,2}$ varies from $-\infty$ to $+\infty$, the corresponding ω_1 in (5.21) form a set. As from (5.11), $g_1(s)$ is continuous on $K_{c,2}$, and the set can be expected to be an interval containing ω_{11}. Thus, this particular interval, denoted by $[\underline{\omega}_1, \overline{\omega}_1]$, is sufficient for the design since it already covers all possible gains. This interval is henceforth referred to as the principal interval of (5.21). Apart from this principal interval, there are possibly many other intervals satisfying (5.21). But the corresponding $g_1(j\omega_1)$ will correspond to a point with the phase of, say, $-3\pi + \varphi_{a,1}$. If such a point is used in design, it may lead to the violation of the Nyquist criterion, although it seems that a given point has been moved to a "safe" position.

In view of the above analysis, only the principal interval $[\underline{\omega}_1, \overline{\omega}_1]$ is needed, although (5.21) may yield more possibilities. The interval is found by starting with $\omega_1 = \omega_{11}$ and increasing or decreasing ω_1 until the equality in (5.21) is achieved. Once the principal interval is determined, (5.20) is solved at each ω_1 in the interval to obtain two real solutions for $K_{c,2}$. It should be pointed that some solution may cause an unstable closed-loop and thus must be discarded. It follows from the Nyquist stability criterion that for closed-loop stability $K_{c,2}$ must be positive if $g_2(0) > 0$, or $K_{c,2}$ has to be negative if $g_2(0) < 0$. Due to the presence of integrators in k_1 and k_2, one sees from (5.12) that

$$g_2(0) = g_{22}(0) - \frac{g_{21}(0)g_{12}(0)}{g_{11}(0)}. \tag{5.23}$$

Thus, the correct solutions for $K_{c,2}$ are those which make $g_2(0)K_{c,2} > 0$ and they are used for the design. Others will be discarded. Using the correct $K_{c,2}$, the corresponding $K_{c,1}$ is readily obtained from (5.19). When ω_1 varies in the principal interval for a given $\omega_2^{(0)}$, the resultant such solutions are denoted by $K_{c,1}(\omega_1, \omega_2^{(0)})$ and $K_{c,2}(\omega_1, \omega_2^{(0)})$. They are plotted in a $K_{c,1}$–$K_{c,2}$ plane as a curve $\mathcal{C}_1$:

$$\mathcal{C}_1 = \left\{ (K_{c,1}(\omega_1, \omega_2^{(0)}), K_{c,2}(\omega_1, \omega_2^{(0)})) \mid \omega_1 \in [\underline{\omega}_1, \overline{\omega}_1] \right\}. \tag{5.24}$$

So far, only (5.17) is considered. A similar discussion can be made to (5.18) and the point of $(K_{c,1}(\omega_1^{(0)}, \omega_2)$, $K_{c,2}(\omega_1^{(0)}, \omega_2))$ varying with ω_2 in the principal interval $[\underline{\omega}_2, \overline{\omega}_2]$ defined similarly to $[\underline{\omega}_1, \overline{\omega}_1]$ will provide the basis for the second curve $\mathcal{C}_2$:

$$\mathcal{C}_2 = \left\{ (K_{c,1}(\omega_1^{(0)}, \omega_2), K_{c,2}(\omega_1^{(0)}, \omega_2)) \mid \omega_2 \in [\underline{\omega}_2, \overline{\omega}_2] \right\}. \tag{5.25}$$

The intersection of the two curves, $\mathcal{C}_1$ and $\mathcal{C}_2$, can be found as

$$(K_{c,1}(\omega_1^*,\omega_2^{(0)}),K_{c,2}(\omega_1^*,\omega_2^{(0)})) = (K_{c,1}(\omega_1^{(0)},\omega_2^*),K_{c,2}(\omega_1^{(0)},\omega_2^*)) \triangleq Z. \tag{5.26}$$

At the intersection Z, the gains are equal, i.e., $K_{c,1}(\omega_1^*,\omega_2^{(0)}) = K_{c,1}(\omega_1^{(0)},\omega_2^*)$ and $K_{c,2}(\omega_1^*,\omega_2^{(0)})) = K_{c,2}(\omega_1^{(0)},\omega_2^*))$. If the arguments are also equal, i.e., $\omega_1^* = \omega_1^{(0)}$ and $\omega_2^* = \omega_2^{(0)}$, then such K_{c,i_1} and $\omega_{i_1}^*$ would be a solution of the simultaneous equations (5.17) and (5.18) and fulfill the design. Unfortunately, it is usually not the case, that is, $\omega_1^* \neq \omega_1^{(0)}$ and $\omega_2^* \neq \omega_2^{(0)}$ in general. To search for a solution, an iteration on ω_1 and ω_2 may be done by substituting ω_1^* for $\omega_1^{(0)}$ and ω_2^* for $\omega_2^{(0)}$ and repeating the above procedure one after another. However, it turns out from extensive simulations that as long as the initial $\omega_1^{(0)}$ and $\omega_2^{(0)}$ are chosen in their principal intervals, the approximate solution in (5.26) with $\omega_1 = \omega_1^*$ and $\omega_2 = \omega_2^*$ can yield satisfactory control performance. Any iterations will generate very little improvement, and thus are not worthwhile carrying out. Therefore, it is recommended that the algorithm begins with $\omega_1^{(0)} = \omega_{11}$ and $\omega_2^{(0)} = \omega_{22}$, and terminates at Z to produce an approximate solution:

$$\omega_1 = \omega_1^*, \quad \omega_2 = \omega_2^*, \tag{5.27a}$$

$$K_{c,1} = K_{c,1}(\omega_1^*,\omega_2^{(0)}), \tag{5.27b}$$

$$K_{c,2} = K_{c,2}(\omega_1^*,\omega_2^{(0)}). \tag{5.27c}$$

The above development can be summarized as the design procedure below, where $g_1(0)$ and ω_{22} are defined respectively by

$$g_1(0) = g_{11}(0) - \frac{g_{12}(0)g_{21}(0)}{g_{22}(0)}, \tag{5.28}$$

and

$$\arg g_{22}(j\omega_{22}) = -\pi + \varphi_{a,2}. \tag{5.29}$$

Multi-loop Modified Ziegler-Nichols Tuning Procedure

Given: $G_p(s)$, type of k_{i_1} (PI or PID) and the specifications (φ_{a,i_1}, φ_{b,i_1} and r_{b,i_1}).

(i) For PI controller, set $\alpha_{i_1} = 0$; For PID controller, set $\alpha_{i_1} = 0.25$ or that from (5.31), or user own choice. Calculate β_{i_1} from (5.16).

(ii) Determine ω_{11} from (5.22). Take $\omega_2^{(0)} = \omega_{22}$ from (5.29), find the principal interval $[\underline{\omega}_1,\overline{\omega}_1]$ of (5.21). Solve (5.20) for $K_{c,2}$ for ω_1 in the principal interval with $K_{c,2}$ satisfying $g_2(0)K_{c,2} > 0$. Obtain the corresponding $K_{c,1}$ from (5.19). Plot the resultant solutions $(K_{c,1}(\omega_1, \omega_2^{(0)}),$ $K_{c,2}(\omega_1, \omega_2^{(0)}))$ in the $K_{c,1}$–$K_{c,2}$ plane to obtain the curve C_1.

(iii) Draw C_2 similarly to (ii).

(iv) Determine the intersection Z of C_1, and C_2 and find ω_{i_1} and K_{c,i_1} from (5.27).

(v) Set $T_{i,i_1} = \beta_{i_1}/\omega_{i_1}$ and $T_{d,i_1} = \alpha_{i_1} T_{i,i_1}$.

In addition to $G_p(s)$, the above procedure has to be supplied with other information on given and desired points and the type of controllers. They should be specified correctly to generate a right and reasonable solution. The detailed instructions and guidelines will be presented below.

Consider first the specification of the given points A_{i_1}. For an SISO process, it is generally accepted that the process critical point provides vital information for controller design and thus it is recommended that the points A_{i_1} be chosen as the critical points of the equivalent processes $g_{i_1}(s)$. Notice that for $g_{i_1}(0) > 0$, the critical point is the place where the Nyquist curve intersects the negative real axis, so that $\varphi_{a,i_1} = 0$. If $g_{i_1}(0) < 0$, it is quite clear that φ_{a,i_1} should equal $-\pi$, instead. Though the above is generally recommended, other points can also be chosen as A_{i_1} if desired. Importantly, the sign of $g_{i_1}(0)$ should still be taken into account in this matter to position right A_{i_1}, as just done with regard to the critical point.

Next, consider the selection of the points B_{i_1} or the choice of φ_{b,i_1} and r_{b,i_1}. It is recommended that the users follow these rules given by (5.6)–(5.9) in Section 5.2. These parameters are given in terms of the process normalized gain, which is unknown for $g_{i_1}(s)$ prior to the multi-loop controller design. It is approximated by:

$$\kappa_{i_1} \approx \left|\frac{g_{i_1}(0)}{g_{i_1 i_1}(j\omega_{c,i_1 i_1})}\right|, \tag{5.30}$$

where $\omega_{c,i_1 i_1}$ is the critical frequency of $g_{i_1 i_1}(s)$ and $g_{i_1}(0)$ is readily evaluated from (5.23) and (5.28).

In practice, a perfect process model does not exist and hence the robustness of the tuning method should be addressed. It is noted that the specification of the desired point (or equivalently φ_{b,i_1} and r_{b,i_1}) can be viewed as a combination of gain and phase margin specifications. Different desired points will yield different closed-loop robustness. In the presence of large modeling errors or severe nonlinearity, it is advisable to choose the desired point which corresponds to large gain and phase margins for the purpose of stable and robust closed-loop performance.

Now the last problem to address is whether to launch a PI or a PID controller for a given process. Generally, a PID controller may give a tighter control of the loop than a PI controller. In case of PID, α_{i_1} is recommended to be:

$$\alpha_{i_1} = \frac{0.413}{3.302\kappa_{i_1} + 1}, \tag{5.31}$$

However, for processes which can be approximated with a first-order model, a PI controller is adequate (Hagglund and Astrom 1991) and derivative action is of little use for these processes. Usually, the derivative term is also turned off when the process has significant noise or a long dead-time.

In the context of multi-loop control, a fairly useful analysis tool for the assessment of the controllability of the process is the Niederlinski index (Luyben 1986). It is shown that a positive Niederlinski index is necessary for the integral stability of the system and a process with negative Niederlinski index will be either wrongly paired or inherently difficult to control by a multi-loop controller. For such processes, the method may not generate satisfactory performance and is thus not recommended for use. It can be easily shown that if the Niederlinski index is positive then the static gain of the equivalent process $g_{i_1}(s)$ and that of the loop diagonal process $g_{i_1 i_1}(s)$ will be of the same sign. As a result, φ_{a,i_1} can be chosen based on the sign of $g_{i_1 i_1}(0)$, instead of the sign of $g_{i_1}(0)$.

5.6 Simulation study

The method is applied to several typical MIMO processes and the performance is compared with the BLT method (Luyben 1986). In all the examples below, the recommendations made in the preceding section are used, and the multi-loop MZN design is illustrated with solid lines and the BLT design with dashed lines. The PID controller is implemented in the form where the derivative action is on the measurement signal to avoid "derivative kick" (Astrom et al., 1993).

Example 5.1:

Consider the well-known Wood/Berry binary distillation column process (Wood and Berry 1973):

$$\begin{bmatrix} y_1(s) \\ y_2(s) \end{bmatrix} = \begin{bmatrix} \frac{12.8e^{-s}}{16.7s+1} & \frac{-18.9e^{-3s}}{21.0s+1} \\ \frac{6.60e^{-7s}}{10.9s+1} & \frac{-19.4e^{-3s}}{14.4s+1} \end{bmatrix} \begin{bmatrix} u_1(s) \\ u_2(S) \end{bmatrix} + \begin{bmatrix} \frac{3.8e^{-8s}}{14.9s+1} \\ \frac{4.9e^{-3s}}{13.2s+1} \end{bmatrix} d(s)$$

Assume that a multi-loop PID controller is adopted. As $g_1(0) > 0$ and $g_2(0) < 0$, $\varphi_{a,1} = 0$ and $\varphi_{a,2} = -\pi$. The normalized gains for the equivalent processes, κ_{i_1} are evaluated from (5.30) as $\kappa_1 = 12.8$ and $\kappa_2 = 4.08$. Thus, it follows that $\varphi_{b,1} = 0.588$, $\varphi_{b,2} = 0.499$, $r_{b,1} = 0.612$ and $r_{b,2} = 0.579$. Using the above data, the multi-loop MZN tuning procedure can be invoked:

Step (i) As PID controllers are used, α_{i_1} are set from (5.31) as $\alpha_1 = 0.0558$ and $\alpha_2 = 0.167$. β_{i_1} are calculated from (5.16) as $\beta_1 = 13.3$ and $\beta_2 = 4.56$.

Step (ii) ω_{11} is determined from (5.22) as $\omega_{11} = 1.61$. Take $\omega_2^{(0)} = \omega_{22} = 0.565$ from (5.29), The principal interval is found as $[\underline{\omega}_1, \overline{\omega}_1] = [1.48, 1.65]$. (5.20) and (5.19) are then solved for ω_1 in $[1.48, 1.65]$ with $K_{c,2}$ satisfying $g_2(0)K_{c,2} > 0$ and C_1 in (5.24) is drawn in Figure 5.2.

Step (iii) Draw C_2 in (5.25) similar to (ii) in Figure 5.2.

Step (iv) The intersection Z of C_1 and C_2 in Figure 5.2 is determined and the solution is found from (5.27) as $\omega_1 = \omega_1^* = 1.643$, $\omega_2 = \omega_2^* = 0.505$, $K_{c,1} = K_{c,1}(\omega_1^*, \omega_2^{(0)}) = 0.949$, and $K_{c,2} = K_{c,2}(\omega_1^*, \omega_2^{(0)}) = -0.131$.

To see that further iterations will generate little improvement, ω_1^* is substituted for $\omega_1^{(0)}$ and ω_2^* for $\omega_2^{(0)}$ and run (ii) and (iii) for one more time. The resultant curves, $C_{i_1}^{\text{New}}$, are also plotted in Figure 5.2. From the figure, it is clear that $C_{i_1}^{\text{New}}$ and C_{i_1} are very close to each other and the new solution found from the intersection Z^{New} of C_1^{New} and C_2^{New} is $\omega_1 = 1.645$, $\omega_2 = 0.504$, $K_{c,1} = 0.940$, and $K_{c,2} = -0.132$, which does not differ significantly from the one obtained without iteration. Therefore a similar performance can be expected from the new solution, and further iterations are not necessary.

Step (v) Calculate T_{i,i_1} and T_{d,i_1} as $T_{i,1} = \beta_1/\omega_1 = 8.09$, $T_{i,2} = \beta_2/\omega_2 = 9.03$, $T_{d,1} = \alpha_1 T_{i,1} = 0.452$ and $T_{d,2} = \alpha_1 T_{i,1} = 1.51$.

The multi-loop PID controller is thus formed as:

$$G_c(s) = \text{diag}\left\{0.949(1 + \frac{1}{8.09s} + 0.452s), \quad -0.131(1 + \frac{1}{9.03s} + 1.51s)\right\}.$$

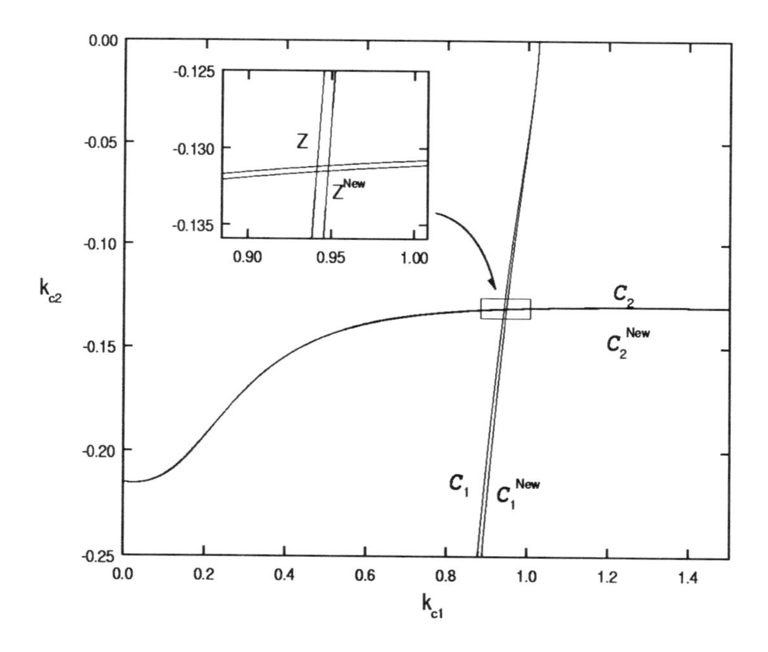

Fig. 5.2. Curves C_1, C_2, $C_1^{\rm New}$ and $C_2^{\rm New}$: Example 5.1

The step responses of the resultant feedback system to a unit set-point change followed by a load disturbance change of -0.5 are shown in Figure 5.3 (solid lines). The corresponding step responses using the BLT method (Luyben 1986) with $K_{c,1} = 0.375$, $T_{i,1} = 8.29$, $K_{c,2} = -0.075$ and $T_{i,2} = 23.6$ is shown in dashed lines in the same figure. It is observed that the multi-loop MZN method gives better loop performance and shorter settling time for the decouplings, especially for the slow (second) loop.

In practice, PI controllers are usually more popular than PID controllers. If a PI controller is adopted, one has $\varphi_{a,1} = 0$, $\varphi_{a,2} = -\pi$, $\varphi_{b,1} = -0.0741$, $\varphi_{b,2} = -0.213$, $r_{b,1} = 0.362$ and $r_{b,2} = 0.369$. The multi-loop MZN method is then activated to yield

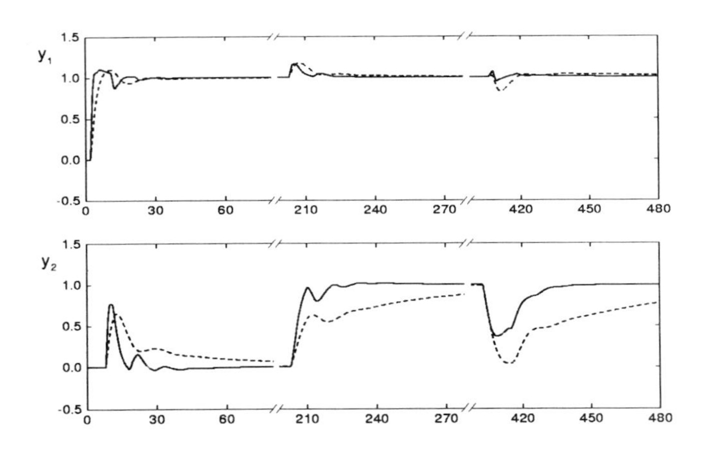

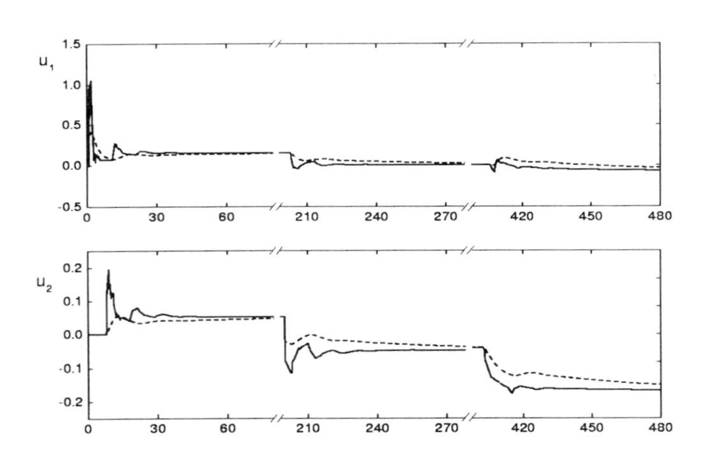

Fig. 5.3. Control performance for PID design: Example 5.1

(— multi-loop MZN; - - - BLT)

Step (i) As PI controller is used, α_{i_1} are set as 0. β_{i_1} are then calculated from (5.16) as $\beta_1 = 13.5$ and $\beta_2 = 4.64$.

Step (ii) $\omega_{11} = 1.61$ and $\omega_{22} = 0.565$ are the same as in the PID case. Take $\omega_2^{(0)} = \omega_{22} = 1.61$, from (5.29), the principal interval is found as $[\underline{\omega}_1, \overline{\omega}_1] = [1.58, 1.79]$. (5.20) and (5.19) is then solved for ω_1 in $[1.58, 1.79]$ and the resulting C_1 in (5.24) is drawn.

Step (iii) Draw $\mathcal{C}_2$ in (5.25) similar to (ii).

Step (iv) The intersection Z of $\mathcal{C}_1$ and $\mathcal{C}_2$ is determined and the solution is found from (5.27) as $\omega_1 = \omega_1^* = 1.60$, $\omega_2 = \omega_2^* = 0.487$, $K_{c,1} = K_{c,1}(\omega_1^*, \omega_2^{(0)}) = 0.699$, and $K_{c,2} = K_{c,2}(\omega_1^*, \omega_2^{(0)}) = -0.0895$.

Step (v) Set $T_{i,1} = \beta_1/\omega_1 = 8.42$, $T_{i,2} = \beta_2/\omega_2 = 9.52$, and $T_{d,1} = T_{d,2} = 0$.

The multi-loop PI controller is thus obtained as:

$$G_c(s) = \text{diag}\left\{0.699(1 + \frac{1}{8.42s}), \quad -0.0895(1 + \frac{1}{9.52s})\right\}.$$

The step responses of the resultant feedback system to the same set-point and load disturbance changes as in the PID case are shown in Figure 5.4 (solid lines) together with the BLT method (dashed lines). Again, it is observed that the multi-loop MZN method gives better loop performance and shorter settling time, especially for the slow (second) loop. Through the comparison of PI and PID performance with the design, one notes that for this example, PI and PID controllers give similar performance, with the PID controller offering slightly faster loop responses.

In order to see the robustness of the closed-loop system, the process is rewritten as $kG_p(s)$, where $G_p(s)$ is the nominal model and k is the scalar gain now deliberately perturbed. The corresponding step responses to the same set-point and load disturbance changes are shown in Figure 5.5, where the response with 20% gain decrease and 20% gain increase are shown with dashed line and dash-dot line respectively. It is noted that the control performance from the perturbed process (dashed lines) is very similar to the ideal one (solid lines).

A real process will have some nonlinearities. When the nonlinearities is modest, the method can be applied directly. For instance, a saturation nonlinearity is introduced into the process $G_p(s)$ such that u_{i_1}, $i_1 = 1, 2$, are limited by $u_{i_1} \in [-0.4, 0.4]$. The corresponding responses are shown in Figure 5.5. It is observed that u_1 does run into saturation, however the control performance from the nonlinear process (dotted lines) is very similar to the linear one (solid lines).

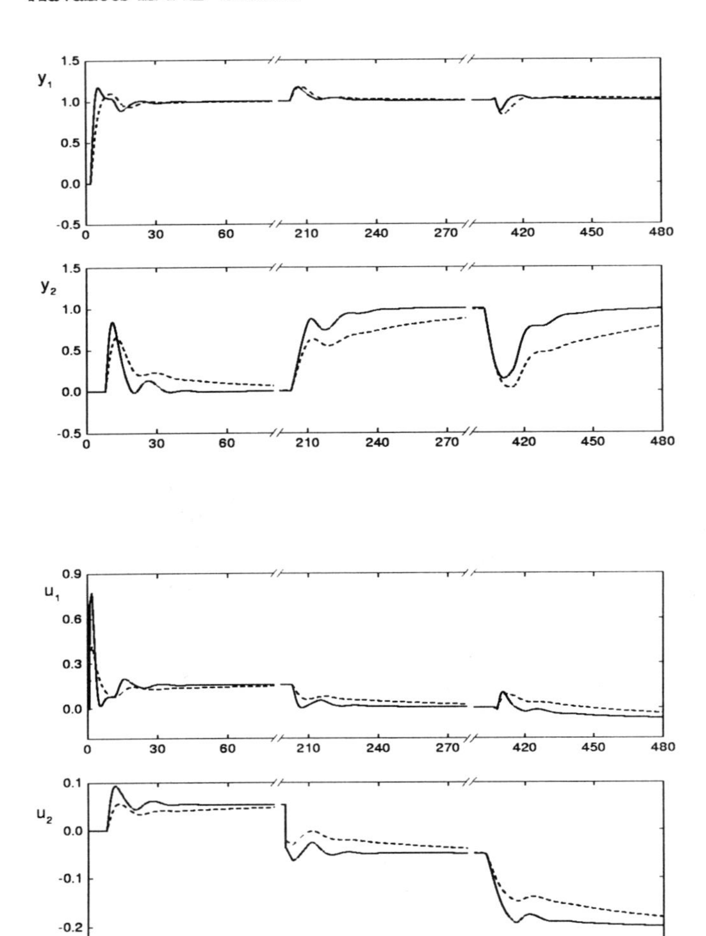

Fig. 5.4. Control performance for PI design: Example 5.1

(— multi-loop MZN; - - - BLT)

Example 5.2:

The process studied by Luyben (Luyben 1986) has the following transfer function matrix:

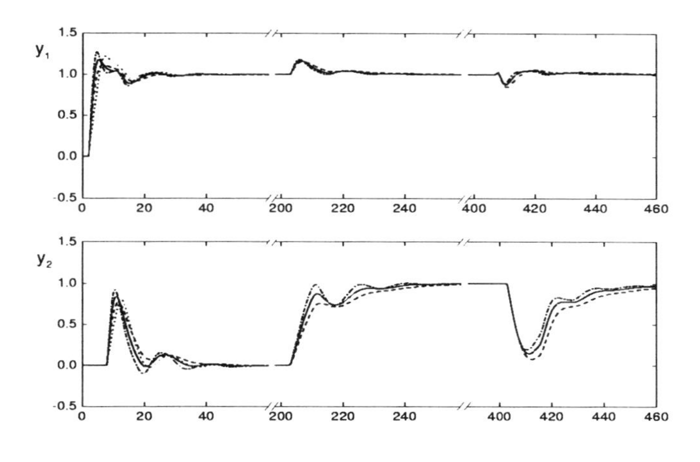

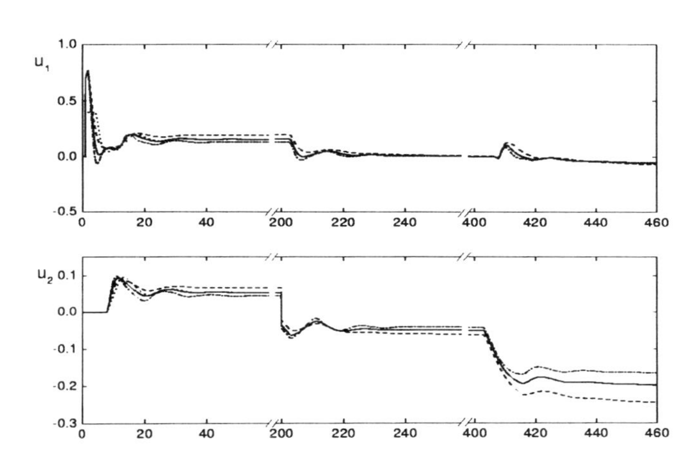

Fig. 5.5. Illustration of robust performance for PI design: Example 5.1 (— ideal; - - - gain decrease; - · - · gain increase; · · · saturation)

$$G_p(s) = \begin{bmatrix} \dfrac{0.126e^{-6s}}{60s+1} & \dfrac{-0.101e^{-12s}}{(45s+1)(48s+1)} \\ \dfrac{0.094e^{-8s}}{38s+1} & \dfrac{-0.12e^{-8s}}{35s+1} \end{bmatrix}.$$

The procedure yields

$$G_c(s) = \text{diag}\left\{68.4(1+\frac{1}{24.8s}+2.85s), \quad -29.5(1+\frac{1}{15.8s}+3.73s)\right\}.$$

The step responses of the resultant feedback system to unit set-point changes followed by load disturbance changes are shown in Figure 5.6 with solid lines. As there is no load disturbance model for this process, load disturbance changes of -30 and 30 are applied directly to the two process inputs, u_1 and u_2, respectively. Here, the magnitude of the load disturbance changes is chosen as 30 to make the load response peak comparable with that caused by the unit set-point changes, so as to exhibit both types of responses clearly in the figure. The corresponding step responses for the BLT method are shown with dashed lines in the figure. Again, the multi-loop MZN controller exhibits better loop and decoupling performance than the BLT method.

Example 5.3:

The 24 tray tower separating methanol and water examined by Vinante and Luyben (Luyben 1986) has the following transfer function matrix:

$$G_p(s) = \begin{bmatrix} \frac{-2.2e^{-s}}{7s+1} & \frac{-1.3e^{-0.3s}}{7s+1} \\ \frac{-2.8e^{-1.8s}}{9.5s+1} & \frac{4.3e^{-0.35s}}{9.2s+1} \end{bmatrix}.$$

Since all its elements are of first-order in nature, the PI controller is used. It is designed with the multi-loop MZN method as:

$$G_c(s) = \text{diag}\left\{-1.46(1+\frac{1}{4.30s}), \; 3.40(1+\frac{1}{5.84s})\right\}.$$

The step responses of the resultant feedback system to unit set-point changes followed by load disturbance changes are shown in Figure 5.7 with solid lines, where load disturbance changes of 0.5 and 1 are applied directly on the two process inputs, u_1 and u_2, respectively. The multi-loop MZN method also gives better loop and decoupling performance than the BLT method(dashed lines).

One can see from all the above examples that with the multi-loop MZN method, the performance enhancement over the BLT method is more noticeable in the slow loops than in the fast ones. Since slow loops dominate the overall system performance, any improvement in their performance is significantly meaningful, and the multi-loop MZN method meets this objective.

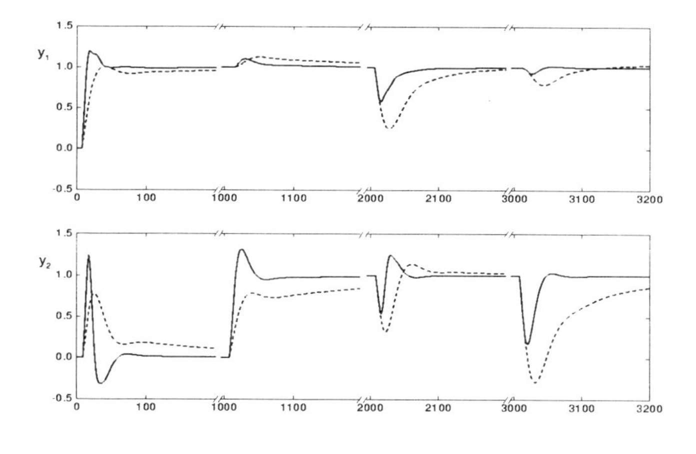

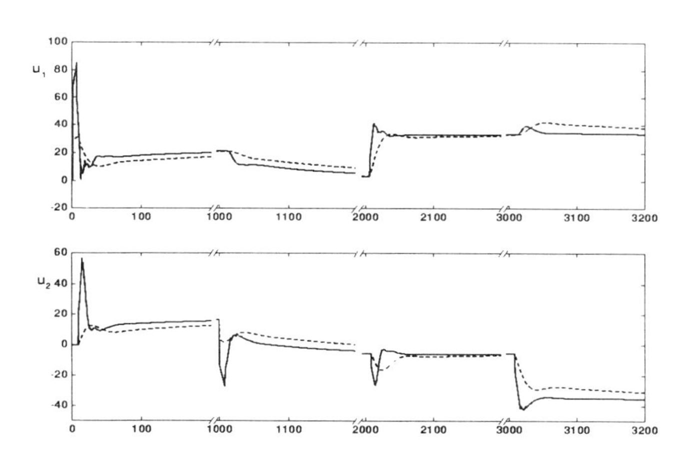

Fig. 5.6. Control performance: Example 5.2

(— multi-loop MZN; - - - BLT)

5.7 Extension to Cross-coupled Controllers

So far, multi-loop controllers without cross-coupled terms have been discussed. Obviously, this type of controllers leaves the process interaction between loops unchanged and is thus suitable only if the interaction is acceptable to users, or it is modest. However, true multivariable controllers with

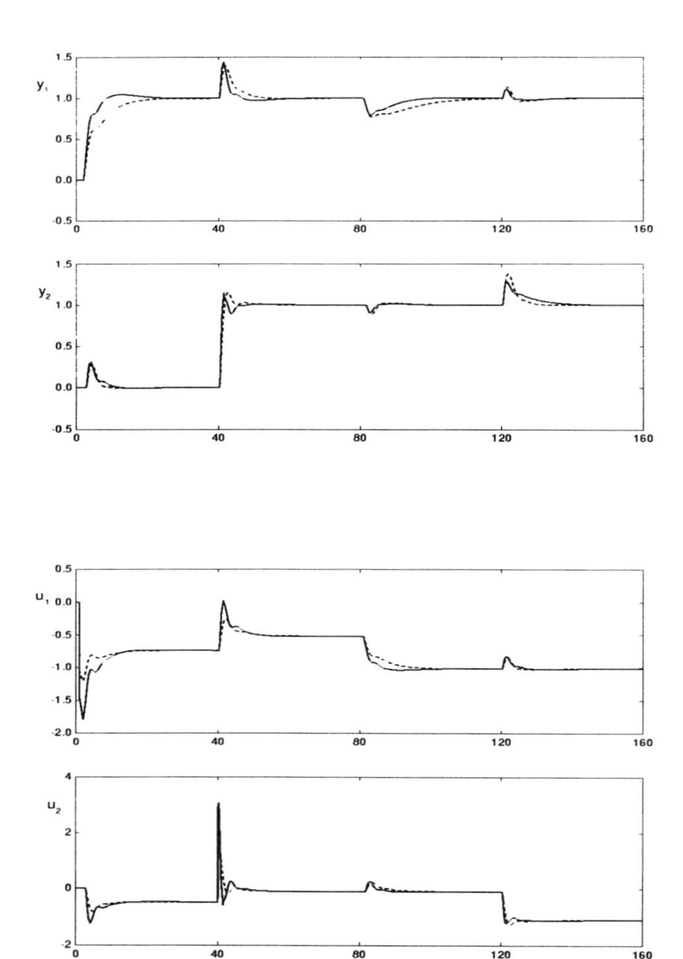

Fig. 5.7. Control performance: Example 5.3

(— multi-loop MZN; - - - BLT)

fully cross-coupled terms are necessary for processes with significant interactions where the performance by multi-loop or decentralized control may be too poor to meet specifications or even the processes may not be stabilized by decentralized control. The easiest way to do multivariable control is use a decoupler to compensate for interactions in conjunction with decentralized control.

A 2×2 multivariable control system is shown in Figure 5.8, where

$$G_d(s) = \begin{bmatrix} 1 & k_{d12}(s) \\ k_{d21}(s) & 1 \end{bmatrix} \tag{5.32}$$

is a decoupler, and

$$G_c(s) = \begin{bmatrix} k_1(s) & 0 \\ 0 & k_2(s) \end{bmatrix}$$

is a usual multi-loop controller.

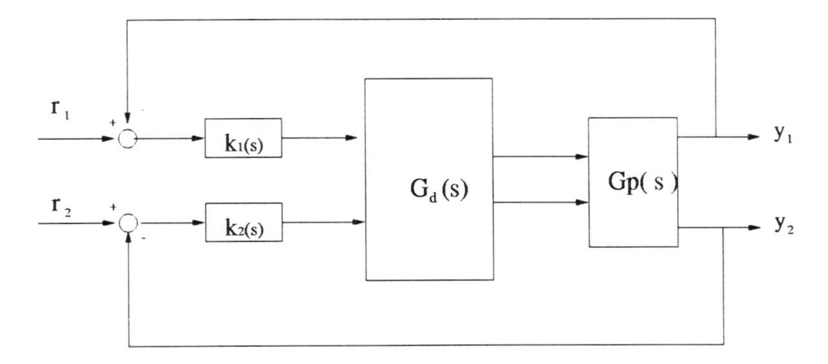

Fig. 5.8. Multivariable decoupling controller

Industrial practice tells that exact decoupling is not realistic since it may lead to an unrealizable or complicated decoupler. What is really needed in practice is that interaction is reduced by a simple decoupler to an acceptable level. To this end, assume that the process is approximately by element-wise first-order plus dead time model:

$$G_p(s) = \begin{bmatrix} g_{11} & g_{12} \\ g_{21} & g_{22} \end{bmatrix} = \begin{bmatrix} \frac{K_{11}}{T_{11}s+1} e^{-L_{11}s} & \frac{K_{12}}{T_{12}s+1} e^{-L_{12}s} \\ \frac{K_{21}}{T_{21}s+1} e^{-L_{21}s} & \frac{K_{22}}{T_{22}s+1} e^{-L_{22}s} \end{bmatrix}. \tag{5.33}$$

One sees that the two off-diagonal elements of the compensated process $Q = G_p G_d$ are

$$q_{21}(s) = \frac{K_{21}}{T_{21}s+1}e^{-L_{21}s} + \frac{K_{22}}{T_{22}s+1}e^{-L_{22}s}k_{d21}(s), \tag{5.34}$$

$$q_{12}(s) = \frac{K_{11}}{T_{11}s+1}e^{-L_{11}s}k_{d12}(s) + \frac{K_{12}}{T_{12}s+1}e^{-L_{12}s}. \tag{5.35}$$

Setting them to zero gives the decoupler elements:

$$k_{d21} = -\frac{K_{21}(T_{22}s+1)}{K_{22}(T_{21}s+1)}e^{-(L_{21}-L_{22})s}, \tag{5.36}$$

$$k_{d12} = -\frac{K_{12}(T_{11}s+1)}{K_{11}(T_{12}s+1)}e^{-(L_{12}-L_{11})s}. \tag{5.37}$$

If $L_{21} \geq L_{22}$, i.e., the channel g_{21} has longer dead time than that of g_{22}, the decoupling route via k_{d21} and g_{22} is fast enough to eliminate the coupled effect of g_{21} on the output y_2 and k_{d21} is realizable by a lead or lag compensator with possible dead time. Otherwise, if $L_{21} < L_{22}$, then k_{d21} in (5.36) is not realizable due to pure prediction term $e^{\alpha s}$, $\alpha > 0$. In this case, one may sacrifice the loop performance for decoupling. K_d in (5.32) then has to be modified as

$$G_d(s) = \begin{bmatrix} e^{-L_{d11}s} & k_{d12}(s) \\ k_{d21}(s) & e^{-L_{d22}s} \end{bmatrix}.$$

It follows that decoupling ($q_{12} = q_{21} = 0$) requires

$$k_{d21} = -\frac{K_{21}(T_{22}s+1)}{K_{22}(T_{21}s+1)}e^{-(L_{d11}+L_{21}-L_{22})s},$$

$$k_{d12} = -\frac{K_{12}(T_{11}s+1)}{K_{11}(T_{12}s+1)}e^{-(L_{d22}+L_{12}-L_{11})s},$$

where

$$L_{d11} = \begin{cases} 0, & if \quad L_{21} \geq L_{22}, \\ L_{22} - L_{21}, & otherwise; \end{cases}$$

$$L_{d22} = \begin{cases} 0, & if \quad L_{12} \geq L_{11}, \\ L_{11} - L_{12}, & otherwise. \end{cases}$$

After the decoupler $G_d(s)$ has been designed in this way, the compensated process $G_p(s)G_d(s)$ has much smaller interaction, and multi-loop controller $G_c(s)$ would now propably be adequate and its tuning may follow the procedures discussed in the previous sections.

Readers may refer to Wang et al. (1997c) for a general and complete approach to multivariable controller design and auto-tuning. In that paper, a method for auto-tuning fully cross-coupled multivariable PID controllers from decentralized relay feedback is proposed for multivariable processes with significant interactions. Multivariable oscillations under decentralized relay feedback are first investigated, and in particular, it is shown that for a stable process the oscillation frequencies will remain almost unchanged under relatively large relay amplitude variations. Therefore, m decentralized relay feedback tests are performed on the process and their oscillation frequencies would be close to each other so that the process frequency response matrix can be estimated at that frequency. A bias is further introduced into the relay to additionally obtain the process steady state matrix. For multivariable controller tuning, a new set of design equations are derived under the decoupling condition where the equivalent diagonal plants are independent of off-diagonal elements of the controller and used to design the controller's diagonal elements first. The PID parameters of the controllers are determined individually by solving these equations at the oscillation and zero frequencies. The applications to various typical processes show significant performance improvement over the existing tuning methods.

CHAPTER 6

PRACTICAL ISSUES

6.1 Introduction

The PID control designs given in Chapters 2 and 3 are based on simple process models which may appear implicitly or explicitly in the control laws. Nominal controller design also assumes the model is perfect and no significant non-idealities exist such as other extraneous signals, disturbances and physical constraints in the system construction. In practice, such ideal conditions rarely hold and special non-trivial effort must be devoted to ensure effective control performance under these situations. Practical non-idealities in the system may arise from various sources:

1. Non-linear elements in the components of the process, including transducers, actuators and the main process itself.
2. Disturbance signals, including high frequency measurement noises and low frequency signal drift due to load disturbances.

This chapter will examine these non-idealities, their effects on PID control systems and some practical ways to cope with them. The chapter will also provide common operational aspects of PID control such as set point weighting for improvement of set point response, automatic-manual bumpless transfer, and important issues in digital implementation of PID controllers.

6.2 Non-linearities

Till now, it has mainly been assumed that the process under control, including the transducers and actuators, is linear. By linearity, it implies that the reaction of the process to changes in the control signal is the same regardless of the part of the working range of the process. The process should thus respond identically, regardless of the size of set point change, disturbance magnitude, or whether the control signal is altered upwards or downwards.

With a linear process, a single controller setting can be used over the entire operational range.

In practice, however, non-linearities always exist in different elements of the process, in the transducers, actuators and the actual process. In many cases when a high level in the control performance is required, the perturbations to the control system may be beyond tolerance threshold and remedial measures are necessary. Compensation may be applied to simple non-linear elements to maintain uniform loop gain. In more severe cases with non-linear elements, compensation may not be applicable and good control performance cannot be attained without modifying the process itself.

6.2.1 Transducer characteristics

Transducers are often non-linear. The reason for this is that the transducers are often not measuring directly the quantity of interest, but rather an indirect quantity which is related to it.

Thermocouples. Temperature transducers made of thermocouples are one example. In the thermocouple, a voltage is measured which is dependent on the temperature. The relationship between voltage and temperature is not linear as given in the following equation.

$$v = a + b\tau + c\tau^2 + \ldots$$

where v is the voltage, τ is the temperature and a, b, c etc., are constants. A calibration table tracking the relationship of the measured and actual variable is often provided. Many controllers are indeed fitted with such a calibration table so that one can linearize these signals. In this way, the controller algorithm really does get a calibrated signal which is proportional to temperature.

pH meters. pH meters are often used for measuring concentrations in a fluid. The pH meter is really measuring the concentration of hydrogen ions in the fluid. The relationship between hydrogen ion concentration and the actual concentration variable is often non-linear. This relationship often looks similar to that shown in Fig. 6.1 with a gain varying according to the working points. If a fixed gain controller is used, it should be tuned according to the highest gain of the process. This means that the controller should be detuned, i.e., given a low gain to prevent instability problems, the consequence being an overall degradation in the response speed of the control system.

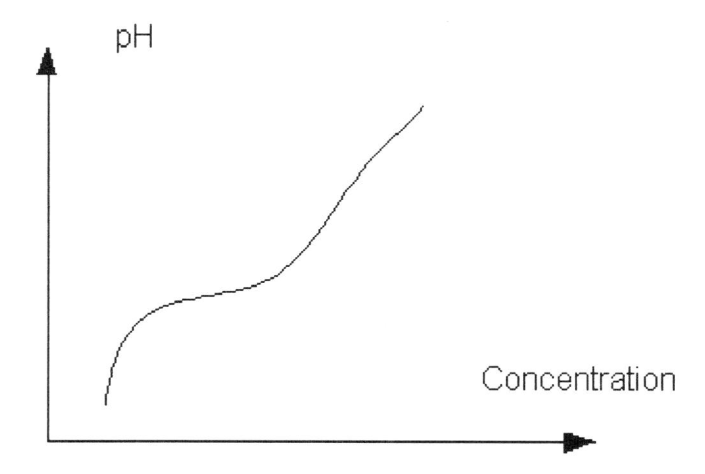

Fig. 6.1. Typical relationship between ion concentration and pH value

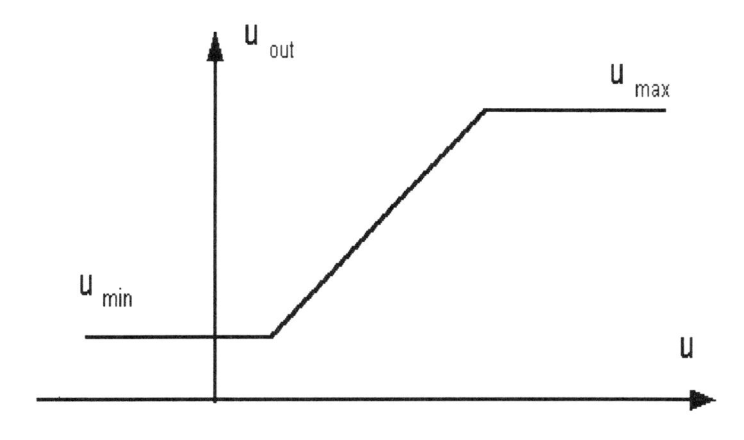

Fig. 6.2. Limited control signal

Flow sensor. Flow is commonly measured by means of a differential pressure sensor. The pressure difference is not a linear function of flow, but the pressure is proportional to the square of the flow rate, given by the following equation:

$$\Delta p = k.q^2,$$

where Δp is the differential pressure, q the volumetric flow and k is a constant.

In order to get a linear process, the square root of the signal has to be taken before it reaches the PID algorithm. The transducers themselves are sometimes provided with a square root algorithm, so that the signals from the device are already linear. In order to cover those cases where the transducers themselves do not linearize the signal, most process controllers have the ability to take the square root of the process variable.

6.2.2 Friction

Friction in control valves is one of the major reasons for poor control loop performance in process control. A valve with high friction causes stick-slip motion and oscillations in the control loop.

Friction is inevitable in mechanical configurations. In all sorts of valves, friction appears in the packing boxes around the valve stem, especially when they are tightened. In ball valves, ball segment valves, and throttle valves there is often also significant friction between the ball/throttle and the seat. The problem is that the static friction mostly increases gradually during operation. The packing boxes are, for example, often tightened after some period of operation in order to avoid leakage.

The friction is varying, both in time and between different operating points. Temperature variations cause friction variations. A high temperature means that the material expands, and therefore that the frictional force increases. Some media give fouling that also increases the friction. Particles in the media may cause damage on the valve. The wear is often non-uniform, so that the friction is different at different valve positions.

If the valve friction has become so high that stick-slip motion occurs, maintenance should be undertaken. The problem is that there are several possible causes of oscillations in control loops, and inappropriate action is often taken. Valve friction is the most likely reason, but it may also be external disturbances or bad controller tuning that causes the oscillations.

Many users assume that oscillations in control loops are caused by bad controller tuning, and therefore they detune the controllers. This is not the appropriate action when valve friction is the cause. Most adaptive controllers behave in the same way. Oscillating disturbances with frequencies in the neighbourhood of the ultimate frequency will detune most adaptive controllers, since the adaptive controllers interpret the oscillations as a result of an excessive loop gain.

6.2.3 Saturation

The signals in control loops are always limited. The measuring transducers have their own working ranges. If the measured quantity falls outside the measurement range of the transducer, the signal will be limited. In the same way, the control signal is also limited, as shown in Fig. 6.2. A valve, for example, has its working range between fully closed and fully opened. The speed of a DC motor is limited to prevent centrifugal forces damaging the motor and couplings. These phenomenons are often referred to as actuator saturation. The effect of saturation is to effectively reduce the gain at high amplitudes, slowing the process response to disturbances.

Limitation of the control signal can cause special problems if the controller is not informed when this is occurring. In controllers, the limitations on the control signals, u_{max} and u_{min}, are often specified and the range should be adhered to as far as possible. Operation outside of the control signal range may risk the controller to a phenomenon known as integral wind-up.

Integrator wind-up. Fig. 6.3 shows the control signal, the process variable and the set point in a case where the control signal is limited. After the first change in set point, the control signal increases up to its upper limit u_{max}. This control signal is not large enough to eliminate the control error. The integral part of the PID controller is proportional to the area under the control error curve. This area is marked in Fig. 6.3. Hence, the integral part will continue to rise since the control signal is unable to eliminate the control error.

Fig. 6.3 shows what happens when after a certain time, the set point is changed to a lower level where the controller is able to eliminate the control error. Because the integral part was allowed to rise and reach a high level while the control signal was limited, the control signal remains at its limit for a longer time before it "unwinds" in accordance to the negative errors.

The problem is known as integrator or reset wind-up and the consequence is a set point response with large overshoot and long settling time. Integrator wind-up commonly occurs in override, surge, batch, and pH control loops. The duration of the windup is longer for those processes with large ultimate periods and large process gains because the corresponding small proportional and integral action prolongs the reversal of controller output.

There are several ways to avoid integrator windup. The simplest way of overcoming the problem is to stop updating the integral part when the control signal is limited (Parr, 1989). For this to be possible, the controller naturally has to know what the limits are. In Fig. 6.4, this point in time corresponds to point B. The question now is at what point it is re-enabled again. Point

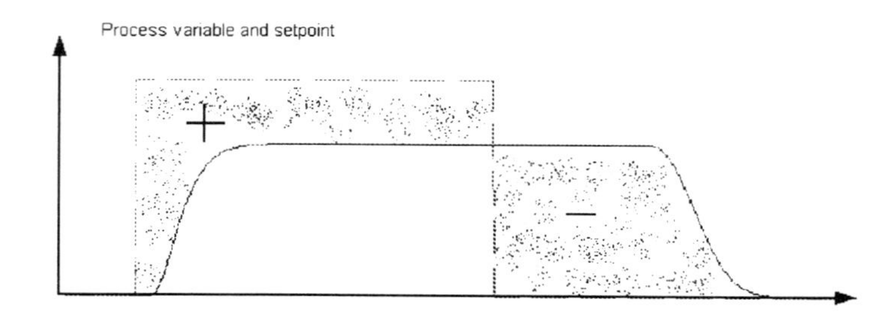

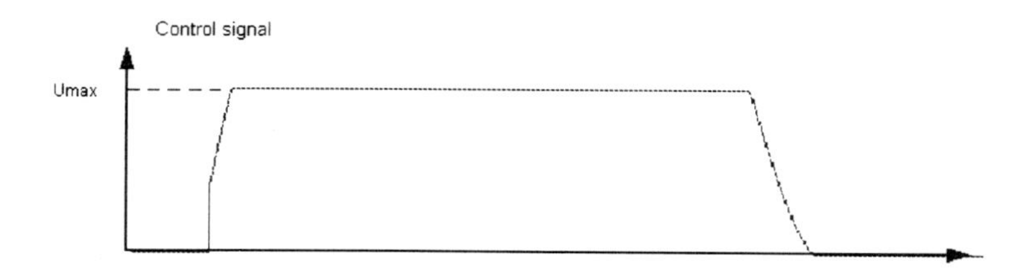

Fig. 6.3. Integrator wind-up

C is obviously far too late although it is still better than point D in the unprotected controller). A common solution is to de-saturate the integral term at the point where the rate of increase of the integral action equals the rate of decrease of the proportional and derivative terms. This occurs when the slope of the PID output is zero, i.e., when

$$e = -T_i \left(\frac{de}{dt} + T_d \frac{d^2 e}{dt^2} \right).$$

This brings the controller out of saturation at the earliest possible moment, but this can, in some cases, be too soon leading to an unnecessarily damped response. Most process controllers are equipped with methods, in one form or another, for avoiding integrator windup.

6.2.4 Hysteresis

Poorly machined gears or loose fitting linkages exhibit backlash. This manifests itself as a differing input/output relationship according to the direction of movement of the input shaft. Stiction and friction cause a similar effect.

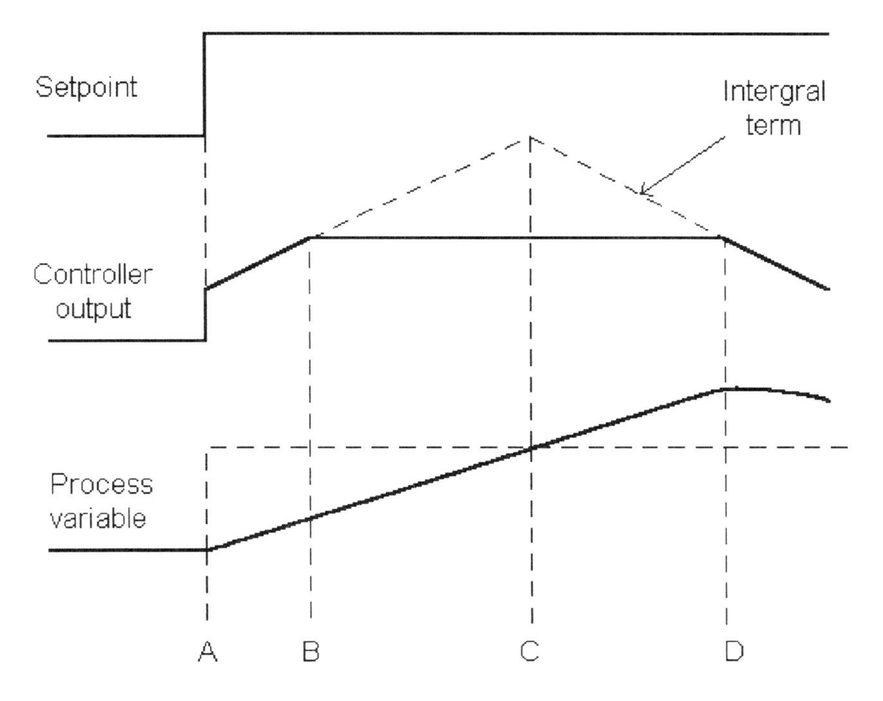

Fig. 6.4. The effect of integral windup

This can be visualized as in Fig. 6.5 and is known in control engineering as hysteresis or backlash.

Hysteresis can be a source of problems and often manifests itself as a "dither" about the set point as the control system hunts in the dead band. It can also be self-reinforcing, as dither will lead to more wear and more backlash.

Hysteresis is best handled by design and careful manufacturing techniques such as spring loaded gears and pre-tensioned linkages.

6.2.5 Dead zone

Dead zone has the response of Fig. 6.6. It is often deliberately introduced into a controller characteristic after the error subtraction to prevent response to small errors. A typical application is a level control system in a surge tank where the level is allowed to vary within limits without corrective action being taken, and it is undesirable for the control system to attempt to correct for ripples and hydraulic resonance.

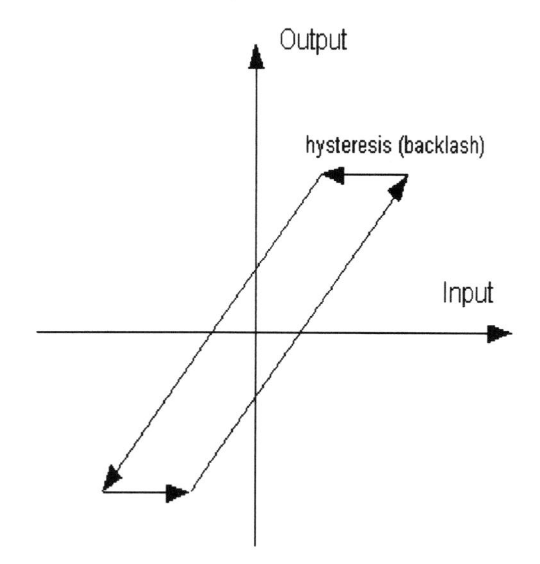

Fig. 6.5. Backlash

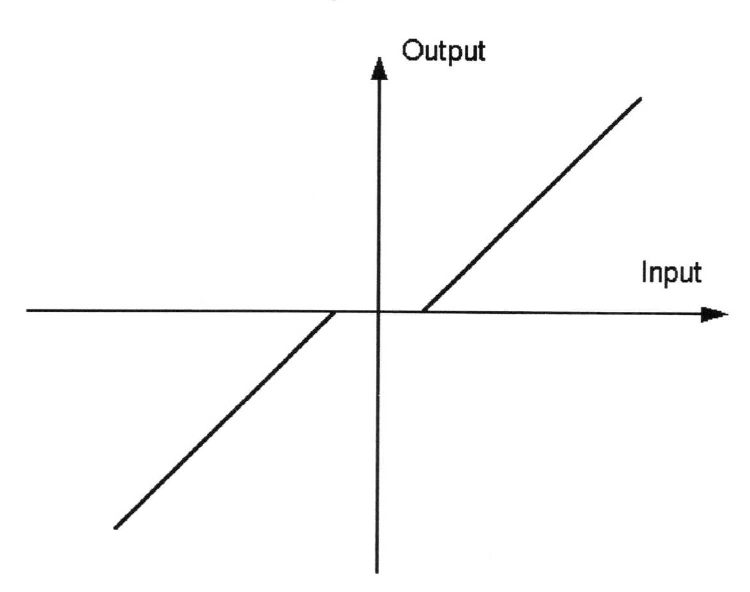

Fig. 6.6. Dead zone

6.2.6 Process characteristics

Processes can also have known non-linearities. The spherical tank of Fig. 6.7 has a volume/depth relationship of Fig. 6.8, and most flow control valves have a distinctly non-linear but known stem position/flow relationship. One solution is to allow the process to only operate over a limited region where it is reasonably linear, i.e., restrict the operational range. Another simple way is to tune the controller for the worst case and accept degraded performance corresponding to a low gain. Yet, another common method is to use gain scheduling for the controller.

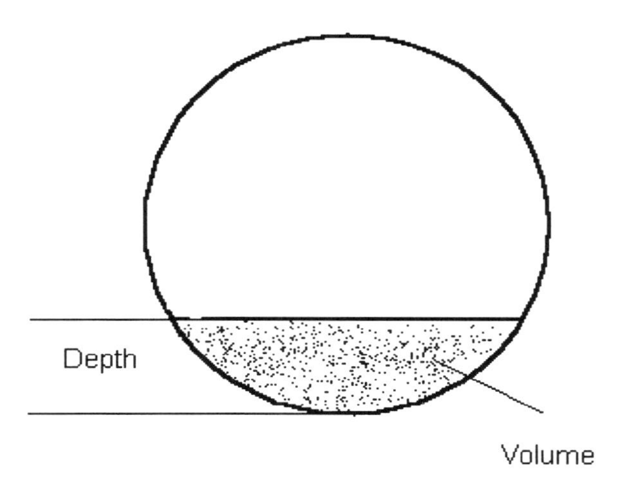

Fig. 6.7. Spherical tank

6.2.7 Gain scheduling

Gain scheduling is an efficient and well proven method to compensate for nonlinearities. The idea is to divide the operating range of the controller into several sub-ranges, and to use different controller parameters within the different sub-ranges.

Suppose, for example, that the process nonlinearity is due to a nonlinear valve. One possibility is to tune the controller for the worst case, which means that the controller is tuned to handle the operating condition where the process gain is large. This may works, but it will typically result in sluggish control in other operating ranges.

Using gain scheduling, a table with a number of sets of controller parameters is built. Automatic tuning methods may be used to build up the table, thus allowing gain scheduling to be more easily used in practice. The controller parameters that are suitable for the actual operating condition are automatically selected from the table based on a suitable signal. For the nonlinear

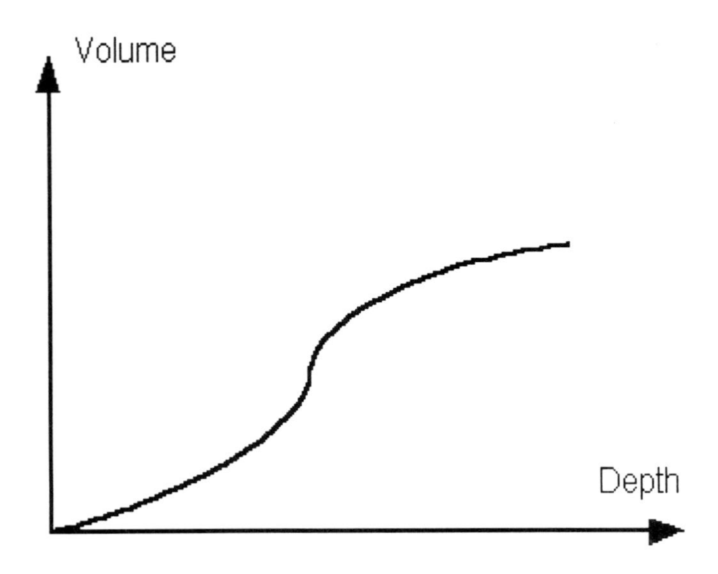

Fig. 6.8. Volume-depth relationship of a spherical tank

valve, this selection is performed using the control signal. It is the control signal that will determine the position of the the valve characteristic, i.e., the gain of the process.

For other types of nonlinearity, other signals are used to determine when to change parameters in the gain scheduling table. Here are some examples of reference signals selected for gain scheduling:

- Nonlinear valves, actuators: The control signal is used to form the reference signal for gain scheduling.
- Nonlinear transducers: The process variable is used to produce the reference signal for gain scheduling.
- Production-dependent variations: Some signal that is related to the production is used as reference for the gain scheduling. It may be the measurement signal, but it may also be some external signal that has to be connected to the controller.

Suppose, for example, that the problem is to control the level in a tank with a variable cross-sectional area. Because of the shape of the tank, the gain of the process varies depending on the level in the tank. In this case, it is desirable to have different controller parameters at different tank levels. The gain scheduling should therefore be controlled using the level signal.

Gain scheduling is often useful for the temperature control of heat exchangers. The temperature controller controls the valve position on the primary side according to the temperature on the secondary side. If the flow on the secondary side varies, this will produce variations in the process dynamics. If, for example, the flow on the secondary side reduces, the medium on the secondary side will stay longer in the heat exchanger, so it will be warmed up for a longer time. This means that the process gain increases. So, in this case gain scheduling can be used with the flow on the secondary side as a reference signal.

In summary, it is important to choose a suitable reference signal for the gain scheduling. One should choose a signal with a clear relation to the actual process nonlinearity.

6.3 Disturbances

Disturbances are extraneous signals in the control systems, often out of the operator's control. They are mostly of three different kinds: disturbances due to set point changes, low frequency load disturbance and high frequency measurement noise. A more exact knowledge of the disturbances may be used to improve the control.

6.3.1 Set point changes

In process control, many control loops operate under a constant set point. A clear exception is the secondary controller in a cascade control structure where the secondary set point is provided by the primary controller. Although the set point may change when the operating conditions need to be varied, it changes typically in steps. Furthermore, step set point changes are usually fed through a low-pass filter so as to smoothen the step change. This is useful because many controllers are tuned for a good load disturbance rejection, but yield poor underdamped set point response.

6.3.2 Low frequency drift

Low frequency disturbances such as slow drift in the load disturbance are compensated for using the controller. The integral part in the PID controller ensures that a high gain is available in these frequency ranges, and therefore effectively eliminates these disturbances.

Disturbances which lie in the same frequency range as the process itself are probably the most difficult to handle. These oscillations cannot be easily filtered out without filtering out other useful process information. In addition, the controller also has difficulty in rejecting these disturbances. If possible, these disturbances should be eliminated from the feedback loop. The best way, if possible, is to eliminate them at the source through proper design of the system and its components. If this is not possible, then feedforward from the source may be an effective method to eliminate their effects before they affect control performance.

6.3.3 High frequency noise

High frequency disturbances at the input to the process are usually controlled effectively by the process itself, because nearly all processes have the characteristics of a low-pass filter. The high frequency components present in the process variables are therefore usually generated in the transducer or the lead from the transducer. These high frequency disturbances or noise should not be tackled at the controller. In order to avoid wear and tear on the valves and actuators, the noise should be filtered out before the process variable is transmitted to the PID algorithm.

If the noise level is so high that the normal filtering in the controller is not sufficient, the process variable signal should be filtered further. In most modern process controllers, the facility is therefore provided for passing the process signal through a low-pass filter. The filter constant should be chosen to be short enough so that the filter does not affect the lower frequency signals in the same band as the controller. The choice of filter constant is again a compromise between speed and stability. With hard filtering (long filter time constant), smooth signals may be obtained with little wear on the valves and actuators as a result. On the other hand, the consequence is a slower handling of load disturbances as the effect of these is also filtered.

Derivative filtering. The derivative part, in the basic form, gives a signal which is proportional to the rate at which the process variable changes. This means that high frequency noise is greatly amplified by the D part. In order to prevent this, it is usual to connect a low-pass filter to the derivative part. The main objective of the low-pass filter is to ensure that derivative part affects signals only within the frequency range of interest. The low-pass filter must therefore have different time constants depending on the dynamics of the process in point. The most common method of solving this problem is by connecting the time constant of the filter to the derivative time:

$$T_{filter} = \frac{T_d}{N},$$

so that the derivative part is given instead by

$$\frac{T_d}{N}\frac{du_d(t)}{dt} + u_d(t) = K_c T_d \frac{de}{dt}.$$

In most products where N is fixed, it lies in the range $5 - 10$. In those cases where the operator himself has to set the value of the filter constant, this range is usually found to be a suitable choice. In a number of cases, it is not the filter time constant which is to be set, but the maximum gain at high frequencies. This gain is equivalent to the numerical value of N.

When measurement noise is significant, the derivative part is really not necessary. In this situation, one can usually manage well with just a PI controller.

6.4 Operational Aspects

6.4.1 Set point weighting

The Ziegler-Nichols tuning methods are popular for tuning PID control. With these methods, PID parameters are tuned to give a good load disturbance response. When the process dead-time is small, the controller settings giving a good load disturbance response will give a set point response which exhibits a large overshoot. One method to deal with this dilemma is to introduce set-point weighting which can reduce the overshoot drastically while the load disturbance response retains its performance.

The PID controller with set point weighting is usually implemented as follows:

$$u = K_c \left((\beta r - y) + \frac{1}{T_i} \int (r - y)dt + T_d \frac{d(r - y)}{dt} \right),$$

where $0 < \beta < 1$ is the set point weighting factor. In many cases, set point r is removed from the derivative portion to avoid the "derivative kick" due to set point changes.

Hang et al. (1992) proposed a variable set point weighting. A large set point weighting factor β_l is used during the initial period of a set point change, in order to maintain a fast rise time. Thereafter, a smaller weighting factor β_s is used to reduce the large overshoot and it is increased to a value β_m in the third period to eliminate the undershoot. The following values of β have been found to work well for many processes encountered in practice with $\Theta = 0 - 0.25$ where Θ is the normalized dead-time defined by $\Theta = \frac{L}{T}$, and L, T are the equivalent dead-time and time constant of the process.

$$\beta_l = 1.1,$$
$$\beta_s = 0.2,$$
$$\beta_m = 0.642\Theta + 0.506.$$

The switching instant is determined by the time at which the error $e = r - y$ reaches a ratio of the initial error $e(0)$ when the set point is changed. In this case, the two switching instants are determined at $e_s = 0.7e(0)$ and $e_m = (-2.055\Theta + 0.195)e(0)$ which divide the transient response into three periods and the values of β_l, β_s and β_m can be used in the respective period. PID settings from the Ziegler-Nichols frequency response method is assumed in these recommendations.

For $\Theta = 0.25 \sim 0.5$, recommended weighting factors are:

$$\beta_l = 1,$$
$$\beta_s = 0.2,$$
$$\beta_m = -14.217\Theta^3 + 14.952\Theta^2 - 4.280\Theta + 0.953,$$

and the switching instants are $e_s = 0.85e(0)$ and $e_m = (4.374\Theta^2 - 4.478\Theta + 0.546)e(0)$ accordingly. PID settings are assumed to be set according to the following formulae:

$$K_c = 0.6k_\pi,$$
$$T_i = (-0.188\Theta + 0.512)T_\pi,$$
$$T_d = 0.25T_i.$$

For $\Theta > 0.5$, the corrected Ziegler-Nichols formulae (e.g., Cohen-Coon formulae) for time-delay processes is deemed to more appropriate and no set point weighting is necessary.

6.4.2 Auto-manual bumpless transfer

The output from the PID algorithm is a function of time and the set point and process variable. When the process is operating in manual mode and if the PID algorithm is in force at the same time, it is highly unlikely that the PID output will be the same as the demanded manual output. In particular, it is more likely that the integral term will cause the output from the PID algorithm to saturate at 0% or 100% output. If no precautions are taken, then when switching from automatic to manual and then back to automatic sometime later, due to the difference in the PID and manual output, a large step change may occur at the controller's output and this will cause a "bump" in the process variable which may not be acceptable in certain cases where high control performance is required.

To avoid this "bump" in the manual mode, the controller output is fed back to the PID algorithm to maintain a PID output equal to the actual manual output by essentially tracking using the integral term. Many controllers are indeed equipped with a control mode and a tracking mode. In the control mode, it operates like a normal controller. In the tracking mode, the integrator matches its output to the input signal. In this case, when it is subsequently switched to the automatic mode, controller's output remains the same. In the same way, in the automatic mode, the manual regulator tracks the controller output, so that again at the instant of transfer to manual, the output will be held at its last value.

This characteristic allows proportional response to set point changes to be inhibited by making those changes with the controller in manual. When it is returned to automatic, integral action alone is applied to the set point; the process variable will then ramp towards its final value without the initial step or "bump".

This balancing signal fed back from the output to the PID algorithm is also used when the controller output is forced to follow an external signal. This is commonly encountered in cascade control systems when the secondary control is on manual or saturated. To prevent the primary controller from an integral windup problem, the primary controller is forced to follow the secondary process variable. This is called the track mode. As before, the PID controller needs to be balanced to avoid a bump when transferring between track mode and automatic mode. The feedback output signal achieves this balance.

6.5 Digital PID Implementation

Most PID controllers today are based on microprocessors or implemented in computer-based distributed control systems (DCS). There are associated consequences which will be briefly addressed here. Comparing microprocessor-based PID controllers with the analog equivalents, there appear to be mainly advantages. The use of a microprocessor means that it has become possible to incorporate more complex functions in the controller such as automatic tuning, alarm handling, filtering, digital control etc. The parameter values can also be tuned more accurately, since the user settings will be precisely the true control settings unlike analog PID where they may be different due to aging of analog components.

There is, however, one important disadvantage. Digital controllers do not process the analog signals directly. They are sampled instead for analysis. Between consecutive samples, the controller gets no information on the intermediate value of the analog signal. This is not a problem as long as the time interval between the samplings is much shorter than the time constant of the controlled process.

Sampling is also effectively equivalent to introducing an additional dead-time into the feedback loop. A sudden disturbance is only detected in the controller after an average time of half a sampling period. The sampled version is thus delayed compared with the actual signal, and the delay is approximately half the sampling period. The selection of the sampling interval is often crucial to the performance of digital control systems.

6.5.1 Selection of sampling interval

The choice of sampling interval is coupled to the time constants of the process. Sampling much faster than the process dynamics leads to data redundancy and relatively small information value in the new data points. Sampling that is considerably slower than the process dynamics leads to serious difficulties in understanding the process response and thus difficulties in controlling it. A rule of the thumb is to choose a sampling frequency about ten times the bandwidth of the process. In the time domain, this corresponds approximately to obtaining about 5-8 samples over the rise time of the process step response. It is thus valuable to obtain the step response of the process before selecting the sampling interval. If prior information of the process cannot be available, it is recommended to sample at the fastest speed the equipment can support.

In this context, it is important to understand how the frequency contents of an analog signal is affected by sampling. According to Shannon sampling theorem, frequencies in the measured signal that are higher than the Nyquist frequency (half the sampling frequency) will be misinterpreted as lower frequency ones. This is known as the aliasing effect. The sampling frequency should thus be chosen to avoid aliasing, otherwise such frequencies have to be filtered out before sampling by using a low-pass filter with cut-off frequency at the Nyquist frequency. Such a filter is called an anti-aliasing or pre-sampling filter.

In certain single loop controllers, the sampling period can be fixed. For example, in process controllers, the sampling interval is usually between 0.1 and 0.3s. In instrument systems, on the other hand, it is often possible to specify the sampling period.

Finally, it should be noted that relay autotuning provides a very useful mean for the selection of sampling rate for PID control automatically. It yields an important ultimate frequency and the sampling frequency can be selected with respect to the ultimate frequency of the process.

6.5.2 Discretization

To implement a continuous-time control law, involving continuous-time signal differentiation and integration such as the PID controller, in a digital computer, it will be necessary to approximate the continuous-time control law by differential equations.

Proportional action. In continuous-time, the proportional term is:

$$u_p(t) = K_c(r(t) - y(t)).$$

This term is implemented simply by replacing the continuous variables with their samples:

$$u_p(kh) = K_c(r(kh) - y(kh)),$$

where h is the sampling interval and kh thus denotes the kth sampling instants.

Integral action. In continuous-time, the integral term is:

$$u_i(t) = \frac{K_c}{T_i} \int_0^t e(\tau) d\tau,$$

where $e(t) = r(t) - y(t)$.

In discrete-time, the integral may be approximated by the area under the $e(t)$ curve from $t = 0$ to $t = kh$ iteratively as:

$$u_i(kh) = u_i((k-1)h) + \frac{K_c}{T_i} A,$$

where A is the area under the $e(t)$ curve from $t = (k-1)h$ to $t = kh$.

As shown in Fig. 6.9, A may be approximated in various ways:

$$A_1 = h.e(kh),$$
$$A_2 = h.e((k-1)h),$$
$$A_3 = h.\frac{e(k-1)h + e(kh)}{2}.$$

The three discretization methods may also be respectively obtained by differentiating the integral equation and applying forward, backward and Tustin's approximation to the differential term.

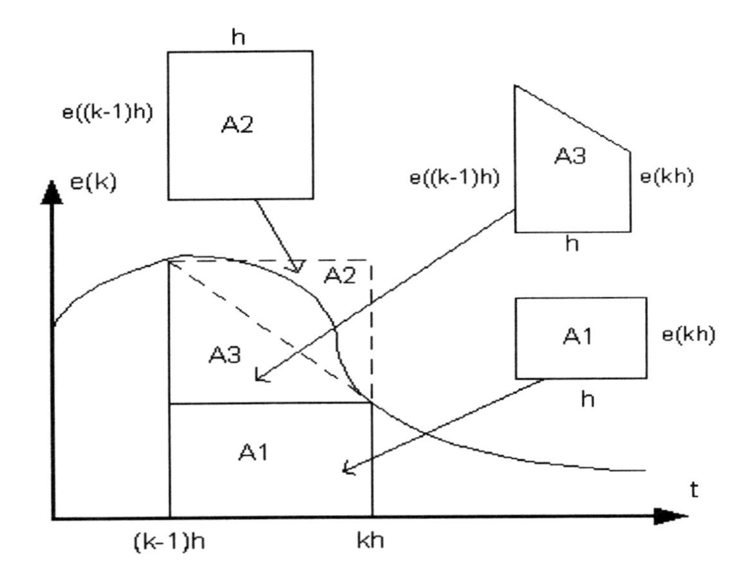

Fig. 6.9. Discretization of PID controller

Derivative action. In continuous-time, the derivative term with filtering is:

$$\frac{T_d}{N}\frac{du_d(t)}{dt} + u_d(t) = K_c T_d \frac{de}{dt}.$$

Integrating this equation, it follows:

$$\frac{T_d}{N}u_d(kh) + \int_0^k h u_d(t)dt = K_c T_d e(kh),$$

$$\frac{T_d}{N}u_d(kh) + \int_0^{(k-1)h} u_d(t)dt + \int_{(k-1)h}^{kh} u_d(t)dt = K_c T_d e(kh),$$

$$\frac{T_d}{N}u_d(kh) + \int_{(k-1)h}^{kh} u_d(t)dt = \frac{T_d}{N}u_d((k-1)h)$$
$$+K_c T_d \left(e(kh) - e((k-1)h)\right),$$

Denoting $A = \int_{(k-1)h}^{kh} u_d(t)dt$, A may be computed in various ways just as for the integral action case:

$$
\begin{aligned}
A_1 &= h.u(kh), \\
A_2 &= h.u((k-1)h), \\
A_3 &= h.\frac{u((k-1)h)+u(kh)}{2}.
\end{aligned}
$$

Thus, the D part can be discretized in the following ways:

$$
\begin{aligned}
u_d(kh) &= \frac{T_d}{hN+T_d}u_d((k-1)h) + \frac{K_cT_dN}{hN+T_d}(e(kh)-e((k-1)h)) \\
u_d(kh) &= \frac{T_d-hN}{T_d}u_d((k-1)h) + \frac{K_cT_dN}{T_d}(e(kh)-e((k-1)h)) \\
u_d(kh) &= \frac{2T_d-hN}{2T_d+hN}u_d((k-1)h) + \frac{2K_cT_dN}{2T_d+hN}(e(kh)-e((k-1)h))
\end{aligned}
$$

These discrete approximations will be stable if the sampling interval h is sufficiently small. For the incremental form of PID control, the control signal is directly obtained by computing the time differences of the controller output:

$$
\Delta u_c(kh) = u_c(kh) - u_c((k-1)h).
$$

APPENDIX A
INDUSTRIAL CONTROLLERS

This appendix is intended to provide a general insight of industrial controllers at the point in time when the book was being written. It is not intended to be exhaustive, focusing on only the representative products of a few manufacturers. Hence, any judgment and conclusion to be drawn from the information presented should be done with caution.

A.1 ABB COMMANDER 351

ABB produces a comprehensive family of process control instrumentation that meets the requirements of a broad range of industries. This product description will focus mainly on the COMMANDER process controllers, in particular the COMMANDER 351 universal process controller.

Control configuration

The COMMANDER 351 universal process controller is a versatile single loop controller. The controller can function in the following modes:

- time proportioning,
- analog PID,
- boundless valve control,
- motorized valve with and without feedback.

The accuracy and error limit of the controllers, referred to nominal range is less than or equal to 0.25%.

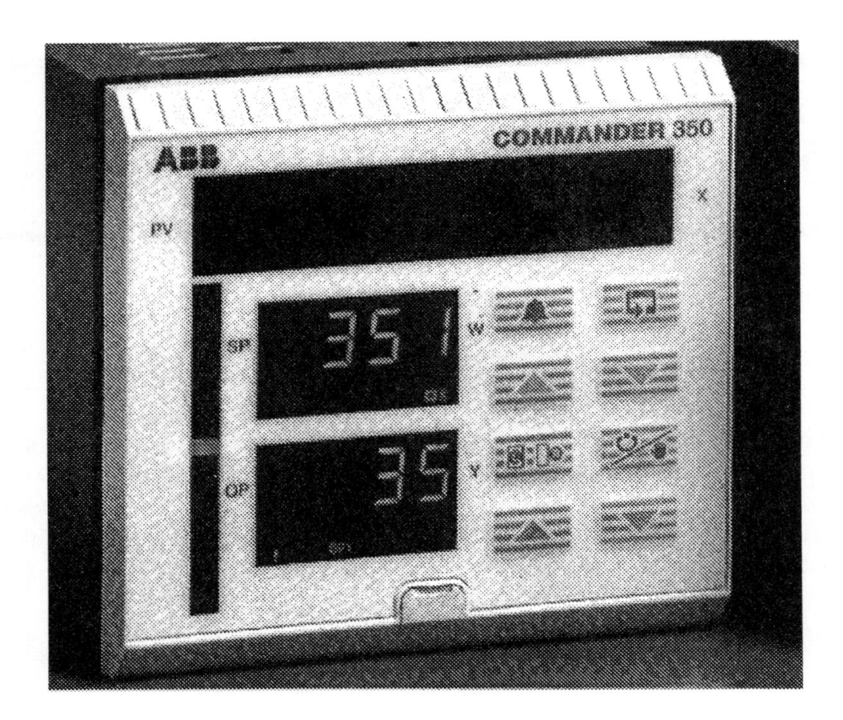

Fig. A.1. ABB COMMANDER 351

Input/output signals

2 universal process inputs are provided and configurable to provide for inputs from thermocouples, resistance thermometers (RTD), and mV, Volts, mA sources.

In addition, 1 fixed analog input for mA and mV connection is provided. Input signals are sampled at 125ms interval for the COMMANDER 351. Digital programmable filter is available for input filtering purposes.

Up to 2 analog outputs are provided. One of them can be programmed as either analog or logic (digital) output. The other is a fixed analog output for a programmable analog range from 0 to 20mA. In addition, 2 standard relay outputs are provided. Each COMMANDER 351 unit has a built-in 2-wire transmitter power supply.

The COMMANDER 351 has 2 separate 15-breakpoint linearizers which can be programmed and applied to either inputs or outputs. These can be used for non-standard thermocouples, nonlinear tank levels or any nonlinear input. The output linearizer accomodates any nonlinear control elements.

Parameter setting

2 autotuning algorithms are provided for on demand control tuning, one to achieve the quarter-damping criterion, the other to achieve minimum overshoot. In addition, *Control Efficiency Monitor* (CEM) measurements are designed to assist in manual fine tuning of the controller. 6 key-performance parameters of the response are measured and displayed, allowing the PID settings to be varied to match the process needs.

For split-range and gain scheduling control purposes, 4 sets of PID settings can be stored, each selectable via digital signals.

Set point programming

Both local and remote set points can be used for the controller. 4 local fixed set points can be stored and they are selectable via digital signals.

Interface

Each unit has a serial interface for parameter setting and configuration via RS485 (2 or 4-wire standard). All COMMANDER products support MODBUS RTU slave protocol and will transmit information over the serial link when interogated by the master.

Functions

4 individual mathematical blocks, each having up to 7 operators and operands, provide functions such as average, maximum and minimum calculations. Square root, relative humidity and arithmetic functions are also included as standard. Inputs can be selected or switched in and out of calculations by digital signals. This allows both simple and advanced calculations to be processed and these can be soft-wired to control functions.

Templates are provided to make the basic configuration for a particular application as simple as possible. When a template is selected, the COMMANDER 351 assumes the preset form for that template. The inputs and software blocks are soft-wired automatically to perform the selected functions.

Additional features

The COMMANDER 351 has intelligent diagnostics and responses which can be used for process safety to initiate an action or to indicate a fault. A processor watchdog monitors the process continuously; a unique loop-break alarm detects analog output failure; and there is an open circuit detector on the input. USing these signals, safety shutdown strategies can be initiated.

The COMMANDER 351 has eight internal process alarms. These can be soft-wired to control strategies, logic equations and output relays. Each alarm can have a separate hysteresis value, programmable in engineering units or time. Alarms can also be enabled or disabled via digital inputs and can be configured as annunciators, so the alarm may be disabled once acknowledged.

Gain scheduling and anti-windup measures are all available with the COMMANDER 351.

Applications

ABB process controllers have typical applications in the following industries:

- environmental,
- food and beverage,
- oil, gas and petrochemical,
- pharmaceuticals,
- power generation and transmission,
- pulp and paper,
- textiles,
- utilities,
- water treatment.

A.2 Elsag Bailey Protonic 500/550

Protonic 500 and Protonic 550 process controllers are Elsag Bailey's flagship all-rounder industrial controllers. They may be used on standalone systems or in a distributed control system network with other controllers, and they may also be linked to higher-level systems.

The two models differ mainly in their front control panels. Protonic 500 has a front panel which distinctly shows the current measured values and operating modes, even from a long distance, in illuminated displays. For operation, all information is clearly presented on an LC display. Protonic 550 has a graphical front control panel with 108x240 dots, thus capable of displaying a large amount of different information.

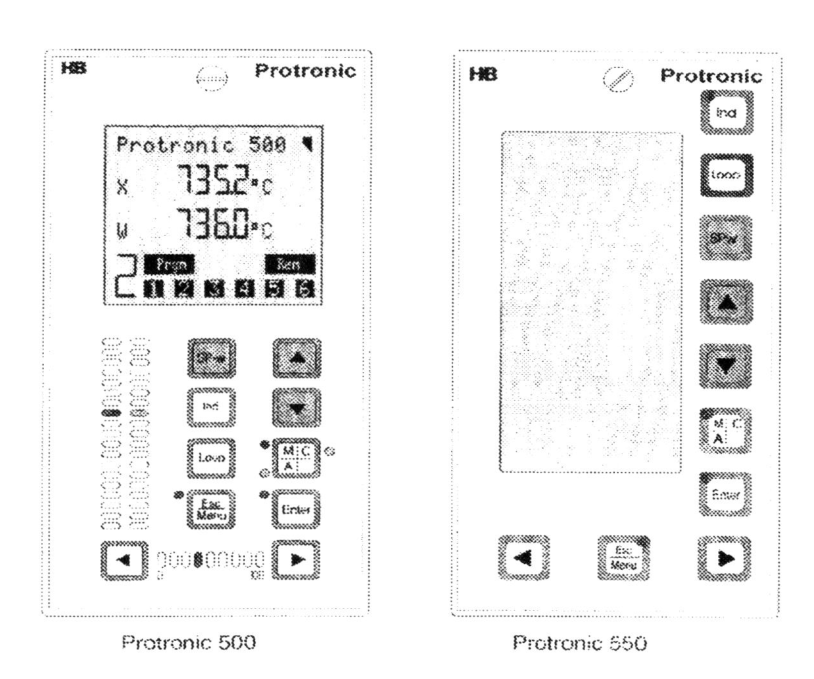

Fig. A.2. Elsag Bailey Protonic 500/550

Control configuration

Protonic 500/550 controllers have 1-4 control channels which may be configured as a fixed-value, ratio, override and cascade structure with selectable P, PI, PD or PID characteristics. The differential action may be configured to act on either the error or process variable. The full PID structure has a series type configuration with the proportional part in series with the integral and derivative portions connected in parallel.

The controller can function in the following modes:

- two-position controller,
- controller for heating/off/cooling, either with two switching outputs or one continuous and one switching output,
- step controller,
- continuous controller, optionally also split-range output with two continuous positioning signals.

In addition, a Smith predictor deadtime compensator is provided for processes with dead-time. The error limit of the controllers, referred to nominal range is less than or equal to 0.2%.

Input/output signals

The basic hardware unit has a universal input to which thermocouples, Pt100 resistance thermometers, and also standard signals 0/4 to 20mA can be connected. When non-linearized temperature transmitters are used, linearization (e.g., square rooting) can be carried out in the controller as the linearization tables for all standard sensors are stored in the unit.

Each unit has an mA input, which is usable as a disturbance variable or set point input. In step controllers, this input can be used for position feedback signal. There is 1 mA output for the positioning signal or other values, e.g., for set point and actual value. For analog input/output signals, sampling interval of 50ms or less can be supported.

4 configurable binary inputs/outputs are optionally usable as controller outputs or alarm value outputs, or as inputs for switchovers in the controller (e.g., manual/automatic).

Expansion slots are available for additional plug-in input/output modules to provide additional inputs/outputs. In addition, a front panel TTL interface is available for connection of a parameter setting and configuration PC.

Parameter setting

Parameters may be entered directly into each channel of controller. Alternatively, a self-tuning mode may be entered where the parameters are extracted from a step test conducted offline in the manual-mode.

Set point programming

Standard ramp function is provided for the set point signal. Each unit also has a configurable programmer which provides a time-dependent set point. Up to 10 programmes with 15 segments each can be stored in the unit. Plug-in PCMCIA memory cards provide memory for saving parameters and user configuration as well as other special algorithms necessary for flow compensation.

Interface

Each unit has a serial (TTL) interface for parameter setting and configuration via a PC. Busable RS-485 interfaces are also available for device fieldbuses such as MODBUS and PROFIBUS, and they also provide connection to higher level systems.

Functions

Pre-defined mathematical functions in the software library enable standard control functions including flow, temperature, pressure control, continuous/On-Off control, Step (valve) control, cascade control, override control, anti-surge control, ratio-control, air/fuel control, combustion control, program control, boiler (drum level) control, and flow compensation. In addition to these preconfigured applications, programming with function blocks according to IEC1131-3 is possible to realize PLC functions and other calculation and control strategies.

Additional features

Each unit is also fully equipped with facilities for anti-windup, signal filtering and conditioning, sensor fault diagnosis and alarms programming.

Applications

Typical applications of Elsag Bailey controllers can be found in:

- packaging machines,
- plastics machines,
- textile machines,
- dryers,
- bakery furnaces,
- small package boiler,
- heat exchangers,
- industrial furnaces,
- air conditioning and ventilation,
- chillers,
- food and beverage,
- laboratories.

A.3 Foxboro 718PL/PR

Foxboro offers different series of process controllers suited for different application requirements. The 731C series of microprocessor-based, 1/4 DIN sized, digital process controllers provide for the basic and standard PID control requirements of process controllers. The 743C is a self-contained, filed-mounted, stand-alone, microprocessor-based controller with an integral power supply. The 743CB is rather similar to the 743C with the ability to perform control functions for up to two independent loops. The 762C controller is a multi-purpose station with the ability to accomplish one or two idenpendent control strategies concurrently. More recently, the I/A Series 1/8 DIN process controllers were launched (718PL, 718PR are general process controllers and 718TC and 718TS are temperature controllers). The following product description will be mainly focused on the 718PL/PR models.

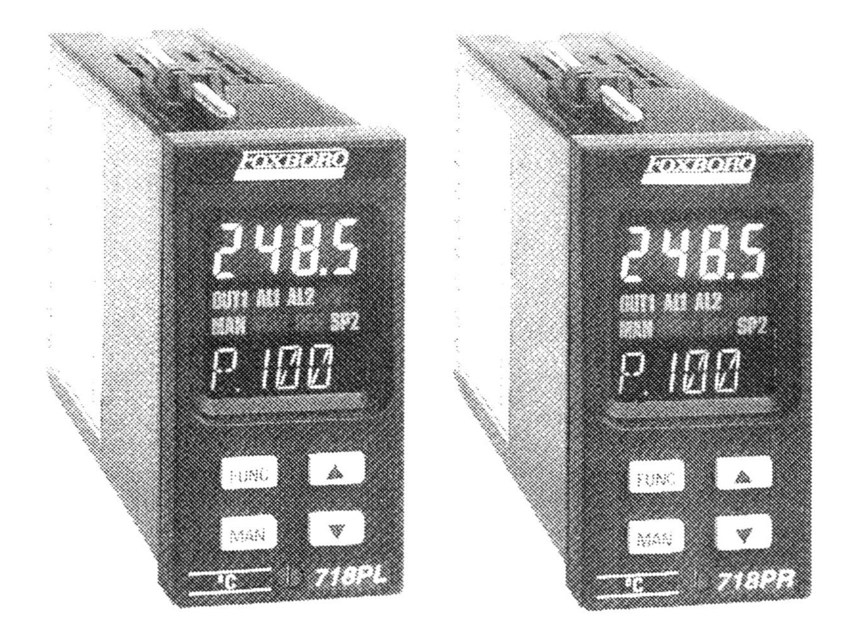

Fig. A.3. Foxboro 718PL/PR

Control configuration

The standard selectable algorithms are P, I, PD, PI, and PID. 4 operational modes are supported:

- continuous control with 0-20 or 4-20 mA output,
- time proportional with relay outputs,
- control with SSR logic voltage drive,
- control with servomotor output.

Control accuracy of ±0.2% of the full scale value is achievable with these controllers.

Input/output signals

Universal inputs are configurable for inputs of the following types:

- thermocouple,
- linear, including mA (0-20mA, 4-20mA), V (0-5V and 1-5V; 0-10V and 2-10V; 0-60mV and 12-60mV).
- RTD.

Outputs can be configured for linear current output (0-20mA, and 4-20mA), relay output,logic voltage output or valve motor drive output. The configurations are fully selectable and keyboard programmable. Signal sampling/updating interval of 150ms is supported on these models. Analog retransmission for 718PR retransmits the linearized and scaled measured value.

The 718PL/PR are equipped with programmable logic inputs. These inputs can be used to:

- enable the remote control from the central unit via the serial link (718PL only),
- start the transfer from main set point to auxiliary set point and vice versa with a programmable ramp,
- select between local and remote set point (718PR only),
- switch between Auto and Manual mode,
- enable/disable the output level limit,
- start the timer function.

Standard and advanced signal conditioning and computation facilities are provided, including signal biasing, filtering, square and square root operation and a multi-segment linearization function.

Parameter setting

Foxboro uses a pattern recognition technique (EXACT/SMART) for the tuning of the controllers. As at the moment, an auto-tuning algorithm automatically calculates the control parameters at start-up. The result is reduced overshoot and good control for normal conditions. The bumpless adaptively tuning algorithm continuously monitor the process and optimize the control parameters during all process phases, including load changes and set point changes. This enables quick startup to be as simple as:

- wiring the instrument,

- configuring set points and alarm threholds,
- initiating the autotune function.

Set point programming

Each unit has a dual set point facility. Set-point change can be driven by the programmable timer or logic inputs. Each set point has a programmable ramp rate and soak time. When the set point is changed, the controller can be configured to follow the programmed ramp rate and soak time. The 718PR model supports remote set point input.

Interface

Foxboro controllers offer two types of protocol, RS-485 with MODBUS or JBUS. Communcation with small systems cxan be done using Foxboro I/A Series system for Windows NT. DDE interface to Windows-based applications provides an integration tool for applications running under Windows.

Functions

Many mathematical functions are provided and integrated into standard control functions such as auto-selector control, cascade control, split range outputs, auto/manual transfer stations, assignable 3-variable indicator functions and user configurable control functions.

Additional features

Anti-windup protection is provided. It provides a configurable band around the set point. When process variable changes or set point changes exceed this limit, the integral term is removed from the controller output.

The 718P series provides Output Limiting, High and Low values, Maximum Rate of Change and Timed Output Limiting. Timed Output Limiting allows configuration of the output limit for a programmable time period. Output limit can be initiated by Contact inputs, Alarms, and Auto/Manual Transfer.

Each instrument are usually equipped with 2 independent alarms. Each alarm can be configured as process alarm, band alarm, or deviation alarm.

Applications

Foxboro controllers are commonly found in:

- continuous kilns for ceramic and building materials,
- polymerization and synthetic fiber plants,
- extrusion and film coextrusion lines; injection presses for plastics,
- rubber vulcanizing plants,
- temperature control in small and medium-sized industrial furnaces,
- lyophilizing and fermentation equipment, reactors for chemical and pharmaceutical industries,
- O.E.M. (Original Equipment Manufacturers),
- packaging and packing equipment,
- food industry,
- environmenal chambers.

A.4 Honeywell UDC3300

Honeywell UDC series of industrial controllers has a wide spectrum of applications. The UDC 3300 is a mid-range general purpose *Universal Digital Controller*. The UDC 5000 and 5300 are higher performance ones, and the UDC 6300 is the top-of-the-line standalone *Process Controller*. This product description will be mainly based on the UDC 3300.

Control configurations

Two independent loops or internally cascaded loops are available in one unit. PID control for independent loop configuration has a parallel structure. Depending on the output algorithms specified (Section 7.4.2), the controller can be configured for the following control algorithms:

- PID.
- PD with manual reset.

Fig. A.4. Honeywell UDC3300

- 3 position step control (Loop 1 only). This algorithm allows the control of a valve (or other actuator), with an electric motor driven by two controller output relays; one to move the motor upscale, the other downscale without a feedback slidewire linked to the motor shaft.
- On-Off (Loop 1 only)

Manual and automatic control modes with local and remote set point, and bumpless, balanceless transfer between modes, are standard features of the controller. High control accuracy of ±0.2% of full scale input span can be achieved. This may be further increased to ±0.05% with field calibration.

Input/output signals

The UDC 3300 is available with one or two universal analog inputs. The first, or *Process Variable* input, can be one of the various thermocouple, RTD, Radiamatic or linear actuations. Linear actuations have thermocouple, RTD, and Radiamatic transmitter characterization capability as a standard feature, and additional linearization features such as the square root capability.

The optional second input is isolated and it accepts the same actuations as the first input, plus it provides the Slidewire input for *Position Proportional Control.* The second input can also be converted into two high level inputs (Input 2 and Input 3). This third input is enabled by first configuring Input 2 as a 4-20 mA or 1-5V type (high level) input. Input 3 will then also be available as a high level input. Input 2 or Input 3 can act as a remote set point or as the second control loop PV input or as a mathematical input. All actuations and characterizations are keyboard configurable. Cold junction compensation is provided for thermocouple type inputs. A configurable single-pole digital filter of 0 to 120 seconds provides input signal damping.

Input/output signal characterizing is achieved via an eight segment characterizer which can be applied to 2 inputs/outputs. It is used for nonlinear control elements. Analog inputs are sampled six times a second.

2 isolated digital inputs are available for remote dry contact closure to select different actions.

The following output types are available, depending on the specific model:

- current output,
- electromechanical relays (5 amps),
- solid state relays (1 amp or 10 amps),
- open collector outputs,
- auxiliary current output.

Parameter setting

Honeywell controllers use ACCUTUNE II for automatic loop tuning. This algorithm, at the touch of a button or through a digital input, accurately identify and tune many processes including those with deadtime and integrating processes. This speeds up and simplifies start-up plus allows retuning at any set point. The tuning is accomplished on-demand using a combination of time domain analysis, frequency response analysis and rule-based expert system techniques.

2 sets of PID gains can be configured for each loop and automatically or keyboard selected. This enables split range control with independent PID tuning constants.

Set point programming

3 local set points can be configured for each control loops. In addition, 6 ramp and 6 soak segments can be stored for set point programming. "Run" or "Hold" of program can be keyboard or remote switch selectable.

Interface

Communications link between the UDC 3300 and host computer or PLC is via the DMCS, RS422/485 ASCII or Modbus RTU communications option.

Functions

2 pre-configured algorithms can be easily implemented into either control loop. They can be linked or used individually and include the capability of using a *Ratio* and *Bias* with any input. The 2 algorithms may be selected from the rich set available: Feedforward summer, summer/subtracter, weighted average, multipler/divider, feedforward multiplier, Hi/Lo select, two characterizers and totalization.

Split-range control is provided with independent PID tuning constants for heating and cooling, plus mixed output forms.

Additional features

Fuzzy logic is used to suppress process variable overshoot due to set point changes or externally induced process disturbances. It operates independently from ACCUTUNE II tuning. It does not change the PID constants, but temporarily modifies the internal controller response to suppress overshoot. This allows more aggressive tuning to co-exist with smooth PV response. It can be enabled or disabled depending on the application or the control criteria.

In addition to autotuning, a gain scheduling feature allows user to schedule eight user defined gain segments over a user defined PV range.

2 fully configurable alarms allow the monitoring of critical process conditions. Alarms can be configured as latching or non-latching. Alarm blocking is also available which allows start-up without alarm energized until after it first reaches the operating region. Alarm can also be set on PV rate of

change, totalizer value and loop break.

Anti-windup is accomplished by setting an output algorithm rate and limit values. Derivative filtering is accomplished with digital filters on inputs. First order lag time constant from 1 to 120 sec can be configured.

Applications

The target industries and applications of Honeywell controllers are mainly:

- manufacturing,
- food and pharmaceutical,
- power generation/industrial utilities,
- chemicals,
- refinery,
- oil and gas,
- pulp and paper.

REFERENCES

Agrawal,A.R., Pandelidis, I.O. and Pecht, M. (1987) Injection-Molding Process Control - A Review, *Polym. Engr. Sci.*, **27**, pp. 1345–1356.

Astrom,K.J. (1982) Ziegler-Nichols auto-tuners. *Internal Report TFRT-3167*. Dept of Automatic Control, Lund Institute of Technology, Lund, Sweden.

Astrom,K.J. (1991) Assessment of achievable performance of simple feedback loops. *Int Journal of Adaptive Control and Signal Processing*, **5**, 3-190.

Astrom,K.J. (1993) Autonomous controllers. *Control Eng. Practice*, **1**(2), 227-232.

Astrom,K.J. and Hagglund, T. (1984a) Automatic tuning of simple regulators. *Proceedings of the 9th IFAC World Congress*, Budapest, 1867–1872.

Astrom, K.J, and Hagglund, T. (1984b) Automatic tuning of simple regulators with specifications on phase and amplitude margins. *Automatica*, **20**(5), 645–651.

Astrom, K.J, and Hagglund, T. (1988a) *Automatic tuning of PID controllers.* Instrument Society of America, NC, USA.

Astrom, K.J, and Hagglund, T. (1988b) A new auto-tuning design. *Preprints IFAC Int. Symposium on Adaptive Control of Chemical Processes, ADCHEM '88,* Lyngby, Denmark, 141–146.

Astrom, K.J, and Wittenmark, B. (1989) *Adaptive Control.* Addison-Wesley: Reading, MA.

Astrom,K.J., Hagglund, T., Hang, C.C. and Ho, W.K. (1992a) Automatic tuning and adaptation for PID controllers - A survey. *Proceedings of the IFAC ACASP '92*, Grenoble, 121–126.

Astrom,K.J., Hang, C.C., Persson, P. and Ho, W.K. (1992b) Towards intelligent PID control. *Automatica*, **28**(1), 1–9.

Astrom,K.J. and Hagglund, T. (1995)*PID Controllers: Theory, Design, and Tuning*, 2nd Edition. Instrument Society of America.

Astrom,K.J., Lee, T.H, Tan, K.K. and Johansson, K.H. (1995) Recent advances in relay feedback methods – a survey. *Proceedings of the IEEE ICMSC'95,* Vancouver.

Astrom,K.J. and Wittenmark, B. (1989) *Adaptive control.* Addison Wesley, Book Company, Reading, Massachusetts, USA.

Babb,M. (1990) Pneumatic instruments gave birth to automatic control. *Control Engineering,* October, pp. 20-26.

Blickley,G. (1988) PID tuning made easy with hand-held computer. *Control Engineering,* November, pp. 99.

Blickley,G. (1990) Modern control started with Ziegler-Nichols tuning. *Control Engineering,* October, pp. 11-17.

Blickley,G. (1990) PID tuning without the math. *Control Engineering,* February, pp. 77.

Bristol,E.H. (1977) Pattern recognition : An alternative to parameter identification in adaptive control. *Automatica,* **13**, pp. 197-202.

Bristol,E.H. (1986) The EXACT pattern recognition adaptive controller, a user-oriented commercial success. In Narendra, Ed., *Adaptive and learning systems,* pp. 149–163, New York, Plenum Press.

Cameron,F. and Seborg, D.E. (1983) A self-tuning controller with a PID structure. *Int. J. Control,* **38**(2), pp. 401-417.

Cartwright,M. (1990) *Fourier Methods for Mathematicians, Scientists and Engineers.* Ellis Horwood.

Chiu, C.P., Shih, M.C. and Wei, J.H. (1991) Dynamic modelling of the molding filling process in an injection moulding machine, *Polymer Engineering and Science*, **31**, pp. 1417–1425

Cohen,G.H. and Coon, G.A. (1953) Theoretical consideration of retarded control. *Trans. ASME,* **75**, pp. 827-834.

Coon,G.A. (1956a) How to find controller settings from process characteristics. *Control Engineering,* **3**, pp. 66-76.

Coon,G.A. (1956b) How to set three-term controller. *Control Engineering,* **3**, pp. 71-76.

Corripio,A.B. (1990) Tuning of industrial control systems. *Instrument Society of America.*

DataTranslation(1995) *Building an Operator Interface in DT VEE, User Manual*; Data Translation, Inc.

Deshpande,P.B. and Ash, R.H. (1981) *Elements of computer process control with advanced control applications,* Instrument Society of America, Research Triangle Park, North Carolina.

Dorato, P., Abdullah, C. and Cerone, V. (1995) *Linear-Quadratic Control: An Introduction*, Prentice Hall, Englewood Cliffs, New Jersey.

Doyle,J.C. and Stein, G. (1981) Multivariable feedback design : concepts for a classical/modern synthesis. *IEEE Trans. Aut. Control,* **AC-26**, 4–16.

Ender,D.B. (1993) Process control performance : Not as good as you think. *Control Engineering,* **40**(10), pp. 180-190.

Eitelberg,E. (1987) A regulating and tracking PI(D) controller. *Int. J. Control,* **45**(1), pp. 91–95.

Fabri,S. and Kadirkamanathan, V. (1996) Dynamic structure neural networks for stable adaptive control of nonlinear systems. *IEEE Trans.on Neural Networks,* **7**(5), pp. 1151–1167.

Fertik,H.A. (1975) Tuning controllers for noisy processes. *ISA Transactions,* **14**, pp. 292-304.

Frank,P.M. (1968) Das PISm - Regelungssystem, eine Erweiterung des PI- Regelungssystems for Totzeitstrecken. *Regelungstechnik,* **16**(7), pp. 306–313.

Franklin, G.F., Powell, J.D. and Naeini, A.E. (1994) *Feedback control of dynamic systems*, (Third Edition), Addison-Wesley.

Friman,M., and Waller, K. (1994) Auto-tuning of Multi-loop Control System. *Ind. Eng. Chem. Res.*, **33**, 1708.

Gawthrop,P.J. (1977) Some interpretation of self-tuning controllers. *Proc. IEE*,**124**, 889.

Gawthrop,P.J. (1986) Self-tuning PID controllers : algorithms and implementation. *IEEE Transactions on Automatic Control*, **31**, pp. 201-209.

Ge, S.Z. (1996) Robust adaptive NN feedback linearization control of nonlinear systems. *Int. Journal of Systems Sciences*, **27**(12), pp. 1327–1338

Gerry,J.P. (1987) A comparison of PID controller algorithms. *Control Engineering,* March, pp. 102-105.

Gerry,J.P. (1988) Find out how good that PID tuning really is. *Control Engineering,* July, pp. 69-71.

Hagglund, T.(1991) *Process control in practice.* Chartwell-Bratt Ltd, Bromley, UK.

Hagglund, T, and Astrom, K.J. (1991) Industrial adaptive controllers based on frequency response techniques. *Automatica,* **27**, 599–609.

Halevi, Y, (1991) Optimal reduced order models with delay. *Proceedings of the 30th Conference on Decision and Control*, Brighton, England, 602–607.

Hang,C.C., Astrom, K.J. and Ho, W.K. (1991) Refinements on the Ziegler-Nichols tuning formula.]em Proc. IEE, Part D,]bf 138(2), 111–118.

Hang, C.C, and Chin, D. (1991) Reduced order process modelling in self-tuning control. *Automatica,* **27**(3), 529–534.

Hang, C.C.,Astrom, K.J. and Ho, W.K. (1993a) Relay auto-tuning in the presence of static load disturbance. *Automatica*, **29**(2), 563–564.

Hang, C.C, Lee, T.H.; Ho, W.K.(1993b) *Adaptive Control*; Instrument Society of America.

Hang, C.C, Wang, Q.G.; Zhou, J.H.(1994) Automatic Process Modelling from Relay Feedback. *Proc. IFAC Symp. on System Ident.*, **2**, 285-290.

Hippe,P., Wurmthaler, C., and Dittrich, F. (1987) Comments on 'A regulating and tracking PI(D) controller', *Int. J. Control,* **45**(5), pp. 1851-1856.

Ho, W.K.,Hang, C.C. and Cao, L.S. (1993) Tuning of PID controllers based on gain and phase margin specifications. *Preprints of the 12th IFAC World Congress,* **5**, 267–270.

Ho, W.K.,Hang, C.C. and Cao, L.S. (1995) Tuning of PID controllers based on gain and phase margin specifications, *Automatica,* **31**(3), pp. 497–502.

Ho, W.K.,Hang, C.C. and Zhou, J.H. (1995) Performance and gain and phase margins of well-known PI tuning formulas, *IEEE Transactions on Control Systems Technology,* **3**(2), pp. 245–248.

Holmberg, U, (1991) Relay feedback of simple systems. *Doctoral Dissertation,* Department of Automatic Control, Lund Institute of Technology.

Honeywell(1995) *Robust Multi-variable Predictive Control Technology, Robust PID.* Honeywell Industrial Automation and Control.

Hornik, K., Stinchcombe, M. and White, H. (1989) Multiilayer feedforward networks are universal approximators. *Neural Networks. 2,* pp. 359–366.

Huang,H.P, Chen, C.L., Lai C.W. and Wang G.B. (1996) Auto-tuning for Model-Based PID Controllers. *AIChE Journal,* **42** (9), pp. 2687-2691.

Jacquot,R.G. (1981) *Modern Digital Control Systems,* Marcel Dekker,inc, New York

Jutan,A. and Rodriguez II, E.S. (1984) Extension of a New Method for online Controller Tuning, *Can. J. Chem. Engng,* **62**, 802-807.

Kaya,A., Akron, O. and Scheib, T.J. (1988) Tuning of PID controls of different structures, *Control Engineering,* July, pp. 62-65.

KentRidge(1992)*Dual Process Simulator KI 100, User Guide;* KentRidge Instruments Pte Ltd.

Kraus,T.W. and Myron, T.J. (1984) Self-tuning PID controller uses pattern recognition approach. *Control Engineering,* 108–111.

Krishnaswamy,P.R., Chan, B.E.M. and Rangaiah, G.P. (1987) Closed-loop Tuning of Process Control Systems.*Chem. Eng. Sci.*, **42**, 2173.

Kuhfittig,P.K.F. (1978) *Introduction To The Laplace Transform.* Plenum Press.

Lam, K.(1982) Design of stochastic discrete time linear optimal regulators -Part I: Relationship between control laws based on a time series approach and a state-space approach. *Int. Journal of Systems Science.* **13**, pp. 979–1000.

Lee, T.H.,Wang, Q.G., Tan, K.K. (1995) A knowledge-based Approach to Dead-time Estimation for Process Control. *Int. J. Control*, **61** (5), 1045-1072.

Lee, T.H.,Wang, Q.G. and Tan, K.K. (1995a) Knowledge-based process identification from relay feedback. *Journal of Process Control*, **5**(6), 387–397.

Leva,A. (1993) PID autotuning algorithm based on relay feedback. *IEE Proceedings D,* **140**(5), pp. 328-338.

Lewis, F.L., Abdullah, C.T. and Dawson, D.M. (1993) *Control of robot manipulators.* Macmillan, New York.

Lewis,F.L. and Syrmos, V.L. (1995) *Optimal Control*, John Wiley & Sons, Inc.

Lewis, F.L., Yesildirek, A. and Liu, K. (1996) Multilayer neural-net robot controller with guaranteed tracking performance. *IEEE Trans. on Neural Networks.* **7**(2), pp. 388–398.

Li, W.,Eskinat, E.; Luyben, W.L.(1991) An Improved Autotune Identification Method. *Ind. Eng. Chem. Res.*, **30**, 1530-1541.

Lin, Y.J.,and Yu C.C. (1993) Automatic Tuning and Gain Scheduling for pH Control. *Chem. Eng. Sci.*, **48**, 3159.

Ljung, L, (1987) *System identification—theory for the user.* Prentice-Hall, Englewoods-Cliff.

Lopez,A.M., Murrill, P.W., and Smith, C.L. (1969) Tuning PI and PID digital controllers. *Instruments and Control Systems,* **42**, February, pp. 89-95.

Lundh, M, (1991) Robust adaptive control. *PhD thesis*, Lund Institute of Technology, Lund, Sweden.

Luyben,W.L. (1986) Simple method for tuning SISO controllers in multivariable systems. *Industrial and Engineering Chemistry: Process Design and Development,* **25**, 654–660.

Luyben, W.L, (1987) Derivation of Transfer Functions for Highly Nonlinear Distillation Columns. *Ind. Eng. Chem. Res.*, **26**, 2490-2495.

Luyben, W.L, (1990) Process Modeling, Simulation and Control for Chemical Engineers; 2nd ed.; *McGraw-Hill: New York.*

Maciehowski,J.M. (1986) *Multivariable feedback design.* Addison-Wesley. Workingham, U.K.

Marlin,T.E. (1995) *Process Control. Designing Processes and Control System for Dynamic Performance*, McGraw-Hill, New York.

Mantz,R.J. and Tacconi, E.J. (1989) Complementary rules to Ziegler and Nichols rules for a regulating and tracking controller. *Int. J. Control,* **49**, pp. 1465–1471.

Marshall,J.E. (1979) Control of time-delay systems. Peter Peregrinus LtD

Marshall,S.A. (1980) The design of reduced order systems. *Int. J. Control,* **30**, 677–690.

Marsili-Libelli,S. (1981) Optimal design of PID regulators. *Int. J. Control,* **49**, pp. 273-277.

Miller,J.A., Lopez, A.M., Smith, C.L., and Murrill, P.W. (1967) A comparison of controller tuning techniques. *Control Engineering,* December, pp. 72-75.

McMillan,G.K. (1983) *Tuning and control loop performance*, Instrument Society of America, Research Triangle Park, North Carolina, second edition.

Morari, M, and Zafiriou, E. (1989) *Robust process control.* Englewood Cliffs, NJ, Prentice Hall.

Murrill,P.W. (1988) *Application concepts of process control.* Instrument Society of America Press, Research Triangle Park, NC.

Nishikawa,Y., Sannomiya, N., Ohta, T. and Tanaka, H. (1984) A method for auto-tuning of PID control parameters. *Automatica,* **20**, pp. 321-332.

Ogunnaike,B.A. and Ray, W.H. (1979)*J. Am. Inst. Chem. Engrs,* **25**, 1043.

Olbrot,A.W. (1981) *IEEE. Trans. Autom. Control,***26**, 513.

Pagano,D. (1991) Intelligent tuning of PID controllers based on production rules system. In *Preprints IFAC International Symposium on Intelligent Tuning and Adaptive Control (ITAC 91),* Singapore.

Palmor, Z.J. and Blau, M. (1994) An auto-tuner for Smith dead-time compensators. *Int. J. Control,* **60**(1), pp. 117–135.

Palmor,Z.J. and Shinnar, R. (1981) Design of advanced process controllers. *AIChe Journal,* **27**(5), 793–805.

Palmor,Z.J., Halevi, Y. and Krasney, N. (1996) Automatic tuning of decentralized PID controllers for MIMO processes. *J. Proc. Control* **42**, 1174–1180.

Parr, E.A. (1989) *Industrial control handbook,* **3**, BSP Professional Books.

Pemberton,T.J. (1972a) PID : The logical control algorithm. *Control Engineering,* May, pp. 66-67.

Pemberton,T.J. (1972b) PID : The logical algorithm - II. *Control Engineering,* July, pp. 61-63.

Persson,P. and Astrom, K.J. (1992) Dominant pole design - A unified view of PID controller tuning. In *Preprints 4th IFAC Symposium on Adaptive Systems in Control and Signal Processing,* Grenoble, France, pp. 127-132.

Polonoyi,M.J.G. (1989) PID controller tuning using standard form optimization. *Control Engineering,* March, pp. 102-106.

Porter,B., Jones, A.H., and McKeown, C.B. (1987) Real-time expert tuners for PI controllers. *IEE Proceedings Part D,* **134**(4), pp. 260-263.

Pintelon,et al. (1994) Parametric Identification Of Transfer Functions In Frequency Domain—A Survey. *IEEE Trans. on Automatic Control,* **39**, 2245-2259.

Radke,F. and Isermann, R. (1987) A parameter-adaptive PID controller with stepwise parameter optimization. *Automatica,* **23**, pp. 449-457.

Rafizadeh,M., Patterson, W.I. and Kamal, M.R. (1996) Physically-based model of thermoplastics injection molding for control applications. *Molding,* **12**, pp. 385–395.

Rafizadeh,M., Patterson, W.I. and Kamal, M.R. (1997) Physically-based adaptive control of cavity pressure in injection molding: filling phase. , *Intern. Polymer Processing,* **12**, pp. 385–395.

Rake,H. (1980) Step Response and Frequency Response Method. *Automatica,* **16**, pp. 519–526.

Ramirez, R.W, (1985) The FFT Fundamentals and Concepts; *Englewood Cliffs, N.J.: Prentice-Hall.*

Rivera, D.E., Morari, M. and Skogestad, S. (1986) Internal model control for PID controller design. *Ind. Eng. Chem. Process Des. Dev.*, **25**, pp. 252–265.

Rotstein,G.E. and Lewin, D.R. (1991) Simple PI and PID tuning for open-loop unstable systems. *Ind. Eng. Chem. Res.*, **30**, pp. 1864–1869.

Rundqwist, L. (1990) Anti-reset windup for PID controllers. In *Preprints 11th IFAC World Congress,* Tallinn, Estonia.

Schei,T.S. (1992) A method for closed-loop automatic tuning of PID controllers. *Automatica,* **28**(3), pp. 587-591.

Seborg, D.E.,Edgar, T.F. and Mellichamp, D.A. (1989) *Process dynamics and control.* John Wiley & Sons Inc.

Shen, S.H.,Wu, J.S.; Yu, C.C. (1996) Use of Biased-Relay Feedback for System Identification. *AICHE Journal.* **42**, 1174-1180.

Shinskey,F.G. (1988) *Process-control systems. Application, design and tuning,* McGraw-Hill, New York, third edition.

Shigemasa,T., Iino, Y. and Kanda, M. (1987) Two degrees of freedom PID auto-tuning controller. *In Proceedings of ISA Annual Conference*, pp. 703-711.

Skogestad,S. and Morari, M. (1989) Robust performance of decentralised control systems by independent designs. *Automatica,* **25**(1), 119–125.

Slotine, J-J.E. and Li, W.P. (1991) *Applied nonlinear control.* Prentice Hall, Englewoods Cliffs, NJ, USA.

Smith, O.J.M, (1957) Closer control of loops with dead-time. *Chemical Engineering Progress,* **53**, 217–219.

Smith,O.J.M. (1959) A controller to overcome dead-time. *ISA J.*, **6**, 28–33.

Smith,C.A. and Corripio, A.B. (1985) *Principles and practice of automatic process control,* New York, Wiley.

Stock,J.T. (1987) Pneumatic process controllers : The ancestry of the proportional-integral-derivative controller. *Trans. of the Newcomen Society,* **59**, pp. 15-29.

Strejc,V. (1980) Least squares parameter estimation. *Automatica,* **16**, pp. 535–550.

Takatsu,H., Kawano, T. and Ichi Kitano, K. (1991) Intelligent self-tuning PID controller. In *Preprints IFAC International Symposium on Intelligent Tuning and Adaptive Control (ITAC 91)*, Singapore.

Tan, K.K.,Wang, Q.G. and Lee, T.H. (1996) Enhanced automatic tuning procedure for PI/PID controllers for process control. *AIChE Journal,* **42**(9), 2555-2562.

Tan, K.K.,Lee, T.H., Lim, S.Y. and Dou, H.F. (1998a) Learning enhanced motion control of permanent magnet linear motor. *Proc. of the third IFAC International Workshop on Motion Control,* Grenoble, France, pp. 397–402.

Tan, K.K.,Lee, T.H., Huang, S.N. and Jiang, X. (1998b) Optimal PI control for time-delay systems based on GPC approach, *Automatica - to appear.*

Tan, K.K.,Lee, T.H., Jiang, X. (1998c) Robust on-line relay automatic tuning of PID control systems. *IEEE Transcation on Control System Technology - to appear .*

Tan, K.K., Lee, T.H., and Huang, S.N. (1999) Adaptive control of ram velocity for the injection molding machine, *Proc.of the 1999 IFAC World Congress*, Beijing, 1999.

Voda,A. and Landau, I.D. (1995) A method for the auto-calibration of PID controllers. *Automatica*, **31**(2)

Wang,Q.G., Hang, C.C. and Bi, Q. (1997a) Process frequency response estimation from relay feedback. *Control Engineering Practice*, **5**(9), pp. 1293–1302.

Wang,Q.G., Hang, C.C. and Bi, Q. (1997b) A technique for frequency response identification from relay feedback. *IEEE Transactions on Control Systems Technology - to appear.*

Wang,Q.G., Zou, B., Lee, T.H. and Bi, Q. (1997c) Auto-tuning of multivariable PID controllers from decentralized relay feedback. *Automatica*, **33**(3), pp.319-330.

Wang,Q.G., Lee, T.H., Fung, H.W. and Bi, Q. (1998) PID tuning for improved performance. *IEEE Transactions on Control Systems Technology - to appear.*

Wang,Q.G., Hang, C.C. and Bi, Q. (1999) A technique for frequency response identification from relay feedback. *IEEE Trans. Control Systems Tech.* **7**(1), pp.122-128.

Wellstead,P.E. (1981) Non-Parametric Methods of System Identification. *Automatica*, **17**, 55-69.

Whitfield,A.H. (1986) Transfer Function Synthesis Using Frequency Response Data. *Int. J. Control*, **43**(5), 1413-1426.

Wolovich,W.A. (1974) *Linear Multivariable Systems.* Berlin: Springer-Verlag.

Wood,R.K. and Berry, M.W. (1973) Terminal composition control of a binary distillation column. *Chem. Engng. Sci.*, **28**, 1707–1717.

Young,P.C. (1970) An instrumental variable method for real time identification of a noisy process. *Automatica*, **6**, pp. 271–287.

Yuwana,M. And Seborg, D.E. (1982) A new method for on-line controller tuning, *AIChE*, **28**, 434-440.

Zhuang, M.and Atherton, D.P. (1991) Tuning PID controllers with integral performance criteria. In *Control 91,* Heriot-Watt University, Edinburg, UK.

Zhuang,M. and Atherton, D.P. (1993) Automatic tuning of optimum PID controllers. *Proc. IEE, Part D.* **140**(3), 216–224.

Ziegler, J.G, and Nichols, N.B. (1942) Optimum settings for automatic controllers. *Trans. ASME,* **64**, 759-768.

Ziegler, J.G, N.B. Nichols,and Rochester, N.Y. (1943) Process lags in automatic-control circuits. *Trans. ASME,* **65**, July, pp. 433-444.

INDEX